Repetitive Project Scheduling

Repetitive Project Scheduling:

Theory and Methods

Li-hui Zhang, Ph.D.
Professor, School of Economics and Management,
North China Electric Power University,
Beijing, China

Xin Zou, Ph.D.
School of Economics and Management,
North China Electric Power University,
Beijing, China

Translator

Li Zhang
Professor, School of Foreign Languages,
North China Electric Power University,
Beijing, China

ELSEVIER

AMSTERDAM • BOSTON • HEIDELBERG • LONDON • NEW YORK • OXFORD
PARIS • SAN DIEGO • SAN FRANCISCO • SINGAPORE • SYDNEY • TOKYO

Elsevier
Radarweg 29, PO Box 211, 1000 AE Amsterdam, Netherlands
The Boulevard, Langford Lane, Kidlington, Oxford OX5 1GB, UK
225 Wyman Street, Waltham, MA 02451, USA

ISBN: 978-0-12-801763-0

British Library Cataloguing-in-Publication Data
A catalogue record for this book is available from the British Library

Library of Congress Cataloging-in-Publication Data
A catalog record for this book is available from the Library of Congress

For Information on all Elsevier publications
visit our website at http://store.elsevier.com/

This book has been manufactured using Print On Demand technology.

CONTENTS

CHAPTER 1

Basic Concept

1.1 PROJECTS

Projects can be defined as temporary rather than permanent social systems, or work systems that are constituted by teams within or across organizations to accomplish particular tasks. "Temporary" means that every project has a definite origin and destination, and "particular" means that the final result of a project cannot be duplicated.

For example, famous projects around the world have included:

- Manhattan Project: developing the first nuclear weapon
- Polaris project: developing a control system for intercontinental missiles

Ubiquitous projects in daily life include:

- Students' homework
- Fashion shows
- Highway construction
- Demonstration

The life cycle of a project may consist of four phases: initiation, planning, execution (including monitoring and controlling), and close-out. A project places emphasis on process, which is a dynamic concept. For example, the construction of a highway could be regarded as a project, but the highway itself cannot be a project.

1.2 REPETITIVE CONSTRUCTION PROJECTS

Repetitive construction projects consist of a set of activities that are repeated in each unit for the length of the job. After an activity is started and/or completed in one unit, it must be repeated in another unit. According to the direction of successive work along the units,

Repetitive Project Scheduling: Theory and Methods.

repetitive construction projects can be divided into two main kinds (Vanhoucke, 2004):

- *Horizontal repetitive projects* are repetitive due to their geometrical layout; among these, highways, tunnels, and pipelines are classical examples. These construction projects are often referred to as *continuous* repetitive projects or *linear* projects due to the linear nature of the geometrical layout and work accomplishment.
- *Vertical repetitive projects.* Rather than a series of activities following each other linearly, vertical repetitive projects involve the repetition of a unit network throughout the project in discrete steps. They are therefore often referred to as *discrete* repetitive projects. Examples are multiple similar houses and high-rise buildings.

Some repetitive construction projects include horizontal repetitive processes and vertical repetitive processes together; Kang et al. (2001) defined these as multiple repetitive projects. A typical example of such projects is multi-story structures.

1.3 CHARACTERISTICS OF REPETITIVE ACTIVITIES AND PROJECTS

As a special kind of project, repetitive construction projects have many characteristics that nonrepetitive projects may not have, such as repetitive and nonrepetitive activities, typical and non-typical activities, resource continuity constraints, distance constraints, and hard and soft logic relations. These characteristics are described below to show the need for a targeted scheduling technique and tool that must be able to model them.

1.3.1 Repetitive and Nonrepetitive Activities

Repetitive activities are those activities that need to be performed in two or more units in the project. On the other hand, nonrepetitive activities are those activities whose sub-activities do not exist in more than one unit. The most common situation is where an activity exists only in the beginning of the project (before starting the first unit) and/or in the first unit. For example, excavation is considered a nonrepetitive activity for high-rise buildings in which it is required only prior to the construction of the first unit (i.e., the first floor). Repetitive construction projects can be made up of all repetitive activities or both repetitive and nonrepetitive activities. Figure 1.1 is an example of a repetitive construction project with nonrepetitive activities; its node network is shown

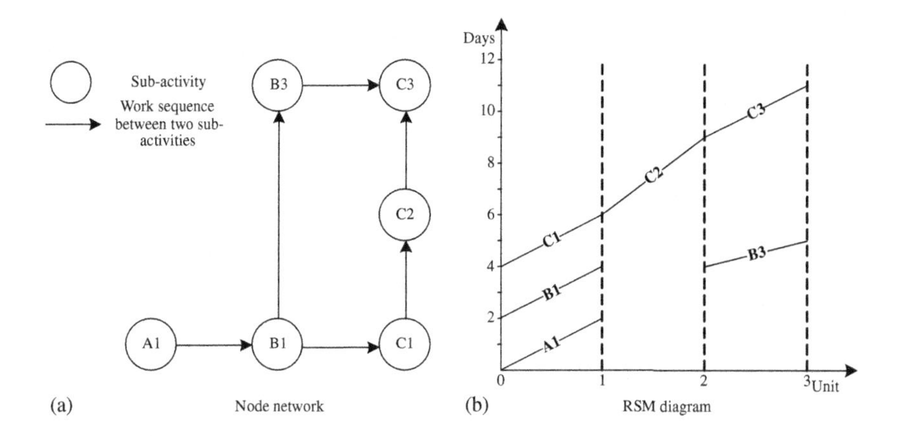

Figure 1.1 A repetitive construction project with repetitive and nonrepetitive activities.

in Figure 1.1(a) where the combinations of capital letters and numbers represent the sub-activities of some activities in some units. For example, "C2" means the sub-activity of activity C in unit 2. By definition, activity A is a nonrepetitive activity, but activities B and C are repetitive activities. The graphical scheduling technique in Figure 1.1(b) is the repetitive scheduling method (RSM), in which the horizontal and vertical axes represent production unit and time, respectively. Sub-activities of an activity in each unit are represented by an oblique line, and each unit is represented by two points: the first denotes the unit start time, and the second denotes its finish time. The vertical difference between the two points is the activity duration for that unit.

1.3.2 Typical and Non-Typical Activities

A typical activity is defined as a series of sub-activities that have the same work amount and duration for each repetitive unit. In contrast, a non-typical activity is a series of sub-activities having different work amounts and, therefore, different durations in different units. If all the activities of a project are typical activities, then the project is a typical project; otherwise it is a non-typical one. Figure 1.2(a) and (b) demonstrate examples of typical and non-typical projects, respectively.

Many scheduling techniques assume that the durations of sub-activities are the same (typical), allowing one to solve the problem easily. However, this assumption is not always practical since activity duration is influenced by many factors such as work amount in each

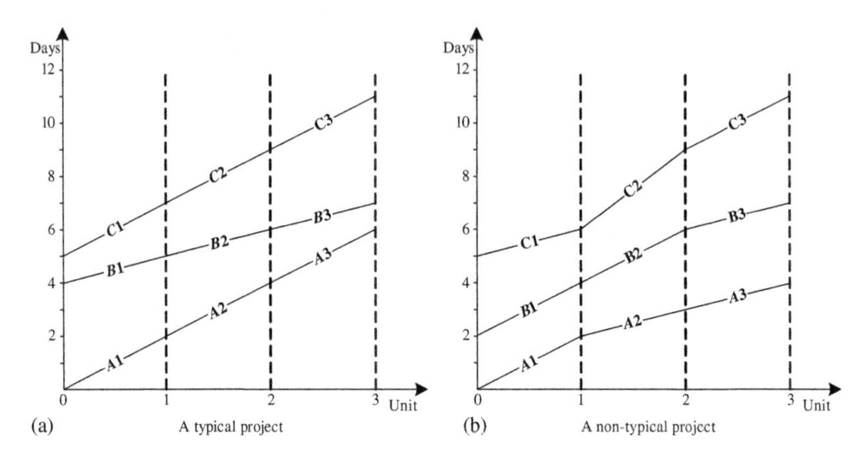

Figure 1.2 Typical and non-typical projects.

unit and resource productivity for each activity. The technique developed should be able to model both typical and non-typical activities.

1.3.3 Resource Continuity Constraints

For each repetitive activity, resource continuity constraints emphasize keeping resources working continuously, without idle time. Idle time is any period that resources are being paid out but not performing any work. Since resources are paid from the date they start working to the date they finish the work, idle time during employment periods is considered unproductive. Accordingly, activities should be scheduled in such a way that idle time of resources is eliminated or minimized. Ensuring resource continuity during scheduling also leads to (1) maximization of the benefits from the learning curve effect for each crew; and (2) minimization of the off-on movement of crews on a project once work has begun. However, Selinger (1980) thought that not all the activities of a repetitive construction project should be required to meet the resource continuity constraint. The author recognizes a trade-off in scheduling repetitive construction projects: work interruption indeed results in an increased direct cost because of the idle time of resources and therefore needs to be avoided. But violation of these resource continuity constraints by allowing work interruption may possibly lead to an overall project duration reduction and corresponding indirect costs, and consequently, a careful trade-off should be made between these two extremes.

A more intuitive comparison is shown in Figure 1.3. The project duration of plan A, in which all the activities must meet the resource

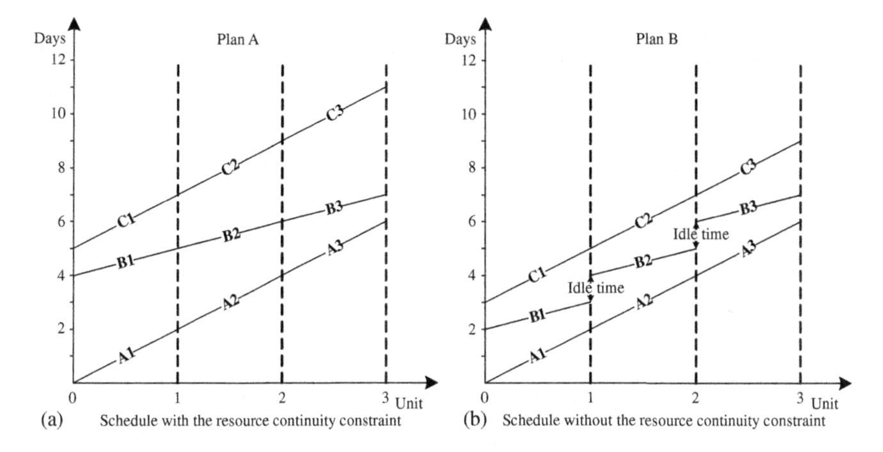

Figure 1.3 Schedule with and without the resource continuity constraint.

continuity constraint (Figure 1.3(a)) is equal to 11 days, and the corresponding resource idle time is zero. On the other hand, in Figure 1.3(b), the project duration of plan B is shortened by 2 days by violating the resource continuity constraint, and it also creates 2 days of idle resource time. Comparing these two plans, both have the same direct project costs, since the durations of all the sub-activities are not changed. At present, if the total cost for covering idle resources under plan B is less than the project indirect cost of plan A, plan B is better than plan A; otherwise, plan B is worse than plan A.

1.3.4 Distance Constraints

Distance constraints are of two types: maximum and minimum distance constraints. The minimum distance constraint indicates that two activities cannot approach each other more than a specified length (or unit) at any time during the project duration. For example, a tunnel's final lining cannot approach excavations more than a specified distance in order to work more effectively and for safety reasons. When planning a vertical repetitive project, the minimum distance constraint is used to ensure resource continuity from one unit/story to the next.

On the other hand, the maximum distance constraint indicates that two activities cannot be further away from each other than a specified distance. An example of such constraint may be "a pipe trench should not be left without being backfilled for more than 500 m for safety reasons." Two activities can be linked with both a minimum and maximum distance constraint.

1.3.5 Hard and Soft Logic Relations

The work sequence between units of an activity is determined by the character of logic relations. In practice, logic relations may be of a "hard" or "soft" character. Hard logic is that inherent in the nature of the work being done. It usually involves technological constraints and often physical limitations (Kallantzis and Lambropoulos, 2004). If the logic relation of an activity is hard, its work sequence between units cannot be changed; for example, the steel structure of a high-rise building must be performed by the fixed sequence from bottom to top.

According to Tamimi and Diekmann (1988), soft logic consists of those relations which allows activities to be scheduled by a variety of work sequences or performed simultaneously in certain circumstances (i.e., the relations are canceled). An example of soft logic in repetitive construction project may be "perform excavation work in four units by the sequence $1 \rightarrow 2 \rightarrow 3 \rightarrow 4$" (assumed by the planner); it is physically possible to "weaken" this relation to generate other optional sequences, for example, "$1 \rightarrow 4 \rightarrow 3 \rightarrow 2$" or "$3 \rightarrow 4 \rightarrow 2 \rightarrow 1$."

In some cases, hard logic is not a good representation of the logical relations of activities, and may unnecessarily limit flexibility in scheduling activities and allocating resources. For example, in Figure 1.4, a housing project consisting of three houses, the sequence of construction for these three houses is not constrained by technological constraints. Therefore, the construction of these houses can be scheduled in many sequences, such as units $1 \rightarrow 2 \rightarrow 3$ as shown in Figure 1.4(b) or units $2 \rightarrow 3 \rightarrow 1$ as shown in Figure 1.4(c). In such a case, constraining repetitive units with hard logic (forcing the sequence of the housing unit $1-3$) would be unnecessary.

Soft logic is the ability of a crew to define its own sequences of units for performing the repetitive work. A comparison of Figure 1.4(b) and (c) shows the benefit of applying soft logic relations to the project. As shown in Figure 1.4(c), reordering the housing units from units $1 \rightarrow 2 \rightarrow 3$ to units $2 \rightarrow 3 \rightarrow 1$ results in a project duration shorter by 2 weeks. Accordingly, the idea of soft logic and its benefits needs to be studied further.

1.4 NETWORK PLANNING TECHNIQUES

Using rational planning and scheduling methods is one key to ensuring the successful completion of a project. Network planning techniques

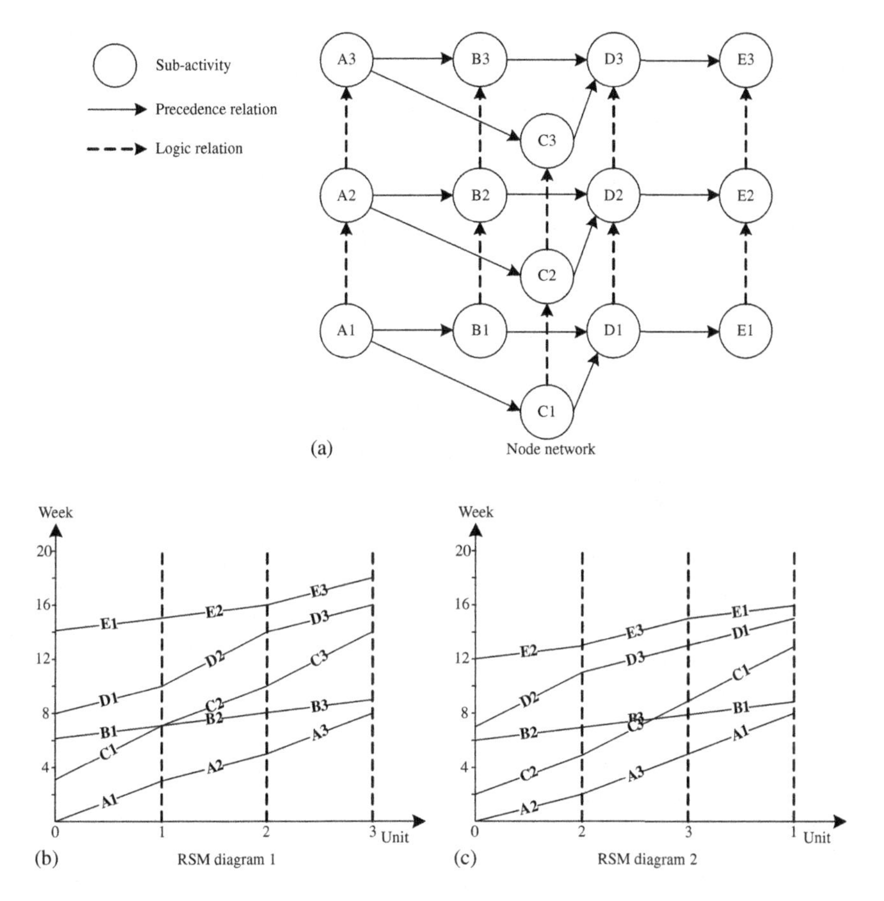

Figure 1.4 A repetitive construction project with soft logic relations.

are methods based on graph theory to analyze, describe, structure, plan, control, and steer projects and processes, whereby time, cost, resources, and other influential factors can be taken into consideration. The two basic network planning techniques are the critical path method (CPM) and the plan evaluation and review technique (PERT).

1.4.1 Critical Path Method

CPM was developed in the 1950s by James Kelly and Morgan Walker (Senior, 1993). The method offers an easy calculation to derive a project schedule and to assess the criticality of activities using the concepts of floats and the critical path, focusing on time. Activities and their precedence relations are depicted in a network by nodes and arrows. Nodes represent activities and activity information such as title, duration, etc.

Arrows represent the precedence relations between activities and the lead time between them. After the network is constructed and the activity durations are given, the calculation of critical path, critical activities, and floats can be performed straightforwardly. The information derived informs project managers of the criticality of activities, which allows them to plan in advance how to schedule the activities and manage the project effectively, based on the current schedule. On the other hand, the managers may decide to alter the original schedule to suit the project deadline, the company resources, and so forth.

Although CPM has been widely used for planning, scheduling, and controlling of construction projects since the late 1950s, it has been recognized as quite unsuitable for repetitive construction projects. The main reasons include:

- It does not guarantee continuity of resources. Although it has been reported by several authors that the uninterrupted utilization of resources is an extremely important issue, neither CPM nor its resource-oriented extensions take these resource continuity constraints into account.
- It does not show the location and time at which a certain crew will be working on a given activity, so it is not efficient for visually monitoring the progress of a particular crew. Moreover, when the distance constraint between activities is violated, CPM cannot provide feedback in time.
- It is believed to be inefficient for large-scale repetitive construction projects, since its calculation becomes tedious and labor intensive (Yang, 2002). For example, a repetitive project consisting of seven activities for 1000 units will require 7000 nodes to represent the network. A network of this size is confusing and unmanageable.

1.4.2 Plan Evaluation and Review Technique

PERT was developed in the late 1950s for the U.S. Navy's Polaris project, which involved thousands of contractors. It has the potential to reduce both the time and cost required to complete a project. A distinguishing feature of PERT is its ability to deal with uncertainly in activity durations. For each activity, the model usually includes three time estimates:

- Optimistic time: generally the shortest time in which the activity can be completed.

- Most likely time: the completion time having the highest probability. This is different from expected time. Seasoned managers have an amazing way of estimating very close to actual data from prior estimation errors.
- Pessimistic time: the longest time that an activity might require.

However, PERT has not been widely used in the construction industry compared to CPM, as it requires more data on activity durations, which is often difficult to obtain or justify. Moreover, PERT requires intensive computation compared to CPM. From the perspective of a repetitive construction project, PERT and CPM have the same limitations due to their underlying time-based scheduling calculation and their graphical presentation in precedence networks.

1.5 EXISTING SCHEDULING TECHNIQUES FOR REPETITIVE CONSTRUCTION PROJECTS

Instead of time-driven techniques (e.g., CPM and PERT), resource-driven techniques have been used to schedule repetitive construction projects such that distance and resource continuity constraints are met and spatial information is shown. These techniques include, but are not limited to:

- Line-of-balance (LOB); see Carr and Meyer (1974), Arditi and Albulak (1986), Al Sarraj (1990), Wang and Huang (1998). LOB is a variation of linear scheduling methods that allows the balancing of operations such that each activity is performed continuously. The major benefit of the LOB methodology is that it provides production rate and duration information in the form of an easily interpreted graphical format. The LOB plot can show at a glance what is wrong with the progress of an activity, and can detect potential future bottlenecks (Arditi et al., 2002).
- Vertical production method (VPM); see O'Brien (1975), Suhail and Neale (1994). VPM is used to schedule the repetitive floors of a high-rise building in conjunction with CPM for non-standard floors. The VPM is essentially a LOB technique tailored to high-rise buildings. Each repetitive floor is modeled as a unit network; the schedule is then created using VPM. The number of crews on a specific activity is adjusted to provide production rates that balance with other activities.

- Horizontal and vertical logic scheduling for multi-story projects; see Thabet and Beliveau (1994). Resources that are considered for scheduling are the physical space requirements of material storage and the movement of manpower and equipment. The scheduling actions proposed to allocate the space resource are: (1) adjustment of productivity rates, (2) interruption of the flow of the activity, and (3) delay in the start of the activity.
- Linear scheduling method (LSM); see Johnston (1981), Chrzanowski and Johnston (1986), Harmelink and Rowings (1998). LSM has long been regarded as a technique that provides significant advantages when applied to linear projects. A linear schedule with time on the horizontal axis and location on the vertical is presented, with activities represented by lines and the slope representing the production rate.
- RSM; see Harris and Ioannou (1998), Zhang and Qi (2012). RSM is similar to LSM; the main difference is that the time constraints in RSM limit the start and finish times of two sub-activities in the same unit by a specified "lag time" or "lead time," but the time constraints in LSM limit the performance time of two activities at the same location by specified maximum or minimum time buffers. Generally speaking, RSM is more adaptable to scheduling discrete projects such as housing projects.

All of the methods described above involve two dimensions: time and location (or unit). They can be classified into two groups: LOB and RSM; the former will be described in detail in Chapter 2, while the latter, as the main planning and scheduling tool in this book, will be introduced in Chapter 3.

CHAPTER 2

Line-of-Balance Technique

2.1 INTRODUCTION

Line-of-balance (LOB) is a technique developed in the 1950s by the U.S. Navy to monitor production-type projects where the delivery of an item is monitored. As a resource-driven technique, the major objective of LOB is to achieve a resource-balanced schedule by determining the suitable crew size and number of crews to employ in each repetitive activity. The major benefit of LOB methodology is that it presents production rate and duration information in an easily interpreted graphical format (Yang and Ioannou, 2004). The LOB plot can present at a glance the progress rate of activities, allowing the possibility to adjust the rates to meet project deadlines, while maintaining work continuity of resources.

Because LOB assumes essentially sequential activities, efforts have been made to combine the benefits of critical path method (CPM) and LOB techniques. One notable effort in this field is the model developed by Suhail and Neale (1994) as a framework for CPM-LOB integration. Their procedure was the first approach to offer a formula for determining crews needed to meet a given deadline. Activities' total float values are used to relax noncritical activities without impacting the project completion date. However, the framework works well only when the calculated number of crews is not rounded to integer numbers. Moreover, it does not consider resource constraints. Since then, more studies have focused on the integration of LOB and CPM methodologies, including studies by Hegazy and Wassef (2001), Ammar and Mohieldin (2002), Hegazy (2002), and Ammar (2013).

In this chapter, a basic definition and graphical representation of LOB will be introduced. Then the successive steps of an integrated CPM-LOB method will be elaborated. Finally, the disadvantages of the LOB technique and future research directions will be discussed.

Repetitive Project Scheduling: Theory and Methods.

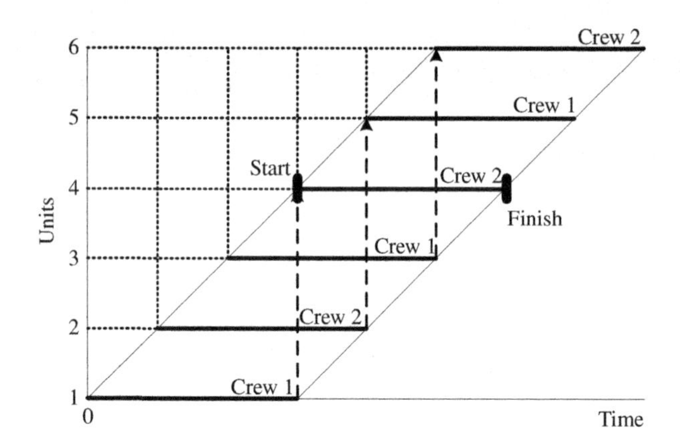

Figure 2.1 Basic representations of repetitive activities.

2.2 BASIC CONCEPT AND REPRESENTATION

The most common representation format of LOB is shown in Figure 2.1, in which each bar represents one activity and each repetitive unit is represented by a horizontal line. The width of the bar is the activity duration of one unit, which is assumed to be uniform across all units. This assumption is not true but it is realistic, especially in projects with a large number of repetitive units. The intersections of a horizontal line at any unit and the activity bar represent the start and finish time for this activity in that unit, respectively.

2.2.1 Crew Synchronization

The representation of LOB allows for multiple crews to be used in the same activity. When several crews are involved in one activity, the LOB schedule assigns tasks for these crews in a regular method, in order to achieve crew synchronization. The specific allocations are: (1) unit 1 is completed by crew 1; and (2) if the work of the jth unit is assigned to crew t, unit $j + 1$ is completed by crew $t + 1$; however, if crew t is the last crew, unit $j + 1$ is completed by crew 1, as shown in Figure 2.1.

2.2.2 Optimum Crew Size and Natural Rhythm

In LOB, the principle of "optimum crew size" assumes that the highest productivity can be achieved as long as an activity is performed in a unit of production by a crew of optimum size. Any crew that is composed of either fewer or more workers is bound to result in lower productivity, as shown in Figure 2.2. The principle of "natural rhythm"

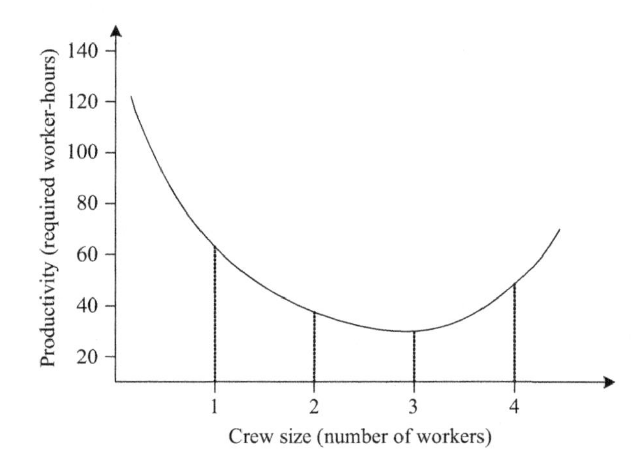

Figure 2.2 The relationship between required worker-hours and crew size.

implies that to increase the progress rate of an activity, multiple crews with optimum size must be employed, such that no idle time occurs when crews move from one unit to another. Consider a U-unit activity; its duration for one unit is D days when the optimum crew size is reached. Then the progress rate of this activity that can meet natural rhythm can only be C/D, where C denotes the number of crews employed.

For a thorough discussion of optimum crew size and natural rhythm, readers are directed to Lumsden (1968), Arditi and Albulak (1986), and Arditi et al. (2002). It may seem difficult to implement the principles of "optimum crew size" and "natural rhythm" in real-life construction projects. However, many construction companies keep records of worker-hours, crew sizes, and daily working hours in previously completed projects. Contractors can estimate the optimum crew size for an activity using this information.

2.3 INTEGRATED CPM-LOB METHOD

The objective of LOB is to achieve a resource-balanced schedule by determining the number of crews to be employed in each repetitive activity. This is conducted such that the units are delivered at a rate that meets a pre-specified deadline and crews' resource continuity is maintained. The analysis also involves determining the start and finish times of all activities in all units and the crews' assignments.

The method consists of three main components, which are discussed in the following subsections.

2.3.1 Meeting a Given Deadline

As shown in Figure 2.3, the end of the project (time T_L) is the date at which the last activity in the last unit is finished. When the first unit of the project is finished, at time T_1, the remaining time until the deadline is taken to complete the remaining $N-1$ units (N is the number of repetitive units). Accordingly, to meet the given deadline, a desired rate of progress (R) can be calculated as follows:

$$R = \frac{N-1}{T_L - T_1} \tag{2.1}$$

where T_L is the deadline of the project and T_1 is the CPM duration of the first unit.

Suhail and Neale (1994) suggested a modification to Eq. (2.1) in order to include noncritical activities such that these activities can be relaxed and given slower rates, taking into consideration the activities' total float. Accordingly, the progress rates were modified as shown in Eq. (2.2).

$$R_i = \frac{N-1}{T_L - T_1 + \mathrm{TF}_i} \tag{2.2}$$

where R_i is the progress rate of activity i, and TF_i is the total float of activity i.

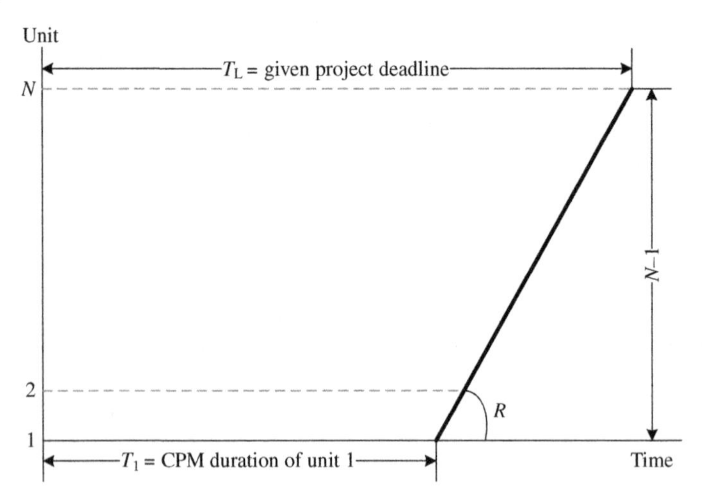

Figure 2.3 Desired project rate of delivery.

The progress rate determined from Eq. (2.2) is, in fact, the minimum progress rate of activity i for meeting the given deadline. In other words, if the progress rate of activity i is less than R_i calculated by Eq. (2.2), then no matter how many crews are employed for other activities, the project cannot be finished within the given deadline.

2.3.2 Number of Crews Determination

When the progress rate is determined, the number of crews (C_i) for the activity can be determined using Eq. (2.3). In general, the number of crews calculated by Eq. (2.3) may not an integer value, and fractional crews are not possible. Therefore, the number of crews must be rounded up to determine the actual number of crews (C_{ai}), as given by Eq. (2.4a). Equation (2.4b) ensures that the actual number of crews allocated to an activity does not exceed crew availability for that activity. Consequently, the actual progress rates (R_{ai}) for different activities must be recalculated, with reference to Figure 2.4, by Eq. (2.5). Equation (2.5) also means that resource continuity is achieved by shifting the start of each unit from its previous one by a time D_i/C_{ai} or $1/R_{ai}$. This shift also has a practical meaning. Because each crew has part of its duration non-shared with other crews, the chances of work delay are reduced when two crews need the same resource, such as a crane.

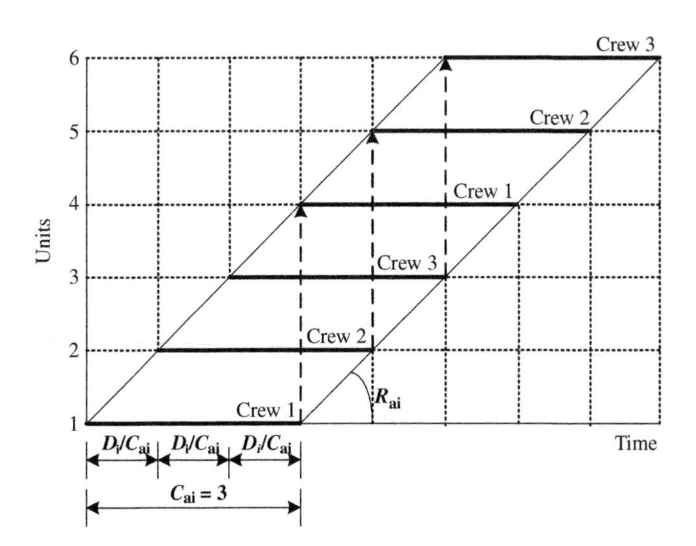

Figure 2.4 Synchronization and resource continuity of crews.

$$C_i = D_i \times R_i \tag{2.3}$$

where D_i is duration of activity i in one unit.

$$C_{ai} = \text{rounded up}(C_i) \tag{2.4a}$$

$$C_{ai} \leq \text{the maximum available crews of activity } i \tag{2.4b}$$

$$R_{ai} = C_{ai}/D_i \quad \text{or} \quad D_i/C_{ai} = 1/R_{ai} \tag{2.5}$$

2.3.3 Drawing LOB Schedule

The resulting LOB schedule becomes simple to draw if all activities run exactly at the desired progress rate R_i without rounding of crews. Otherwise, those activities which need to round up the number of crews will see a greater progress rate than in a theoretical sense. This may lead to delay of the project if resource continuity is to be maintained. In this case, the original schedule will need to be amended. A simple approach is to reschedule the project with a deadline that is slightly shorter than originally desired. In general, however, redrawing the schedule should be done carefully.

In working out the LOB schedule using the actual rate of progress of activities, it is necessary to comply with the precedence relations among activities. When an activity is considered, its predecessors are examined first to identify their latest finish times, which are then considered as a boundary on the start of the current activity. Also, in terms of presentation, showing all the activities on the same chart results in a crowded schedule and can be confusing, even for a small network. To solve this problem, a feasible method is to draw the critical paths in one chart and show the other noncritical paths in another chart. The benefit of drawing these paths is to help visualize the successor and predecessor relations for any given task and accordingly facilitate any desired changes to rates or crews. However, this method has a significant disadvantage: it does not apply to large-scale projects, and when the number of crews on an activity changes, it will be harder to update all charts. Once the schedule is drawn, the start and finish times for each unit in each activity can be read and crew assignments shown.

2.3.4 Example Application

Steps and features of the integrated CPM-LOB method are demonstrated by an example application with 10 identical units. The desired contract duration is 40 days and a minimum buffer time of one day is

Table 2.1 Date for the Example Application

Activity Number	Description	Duration (days)	Preceding Activities
1	Locate and clear	1	–
2	Excavate	3	1
3	String pipe	1	1
4	Lay pipe	4	2,3
5	Pressure test	1	4
6	Backfill	2	5

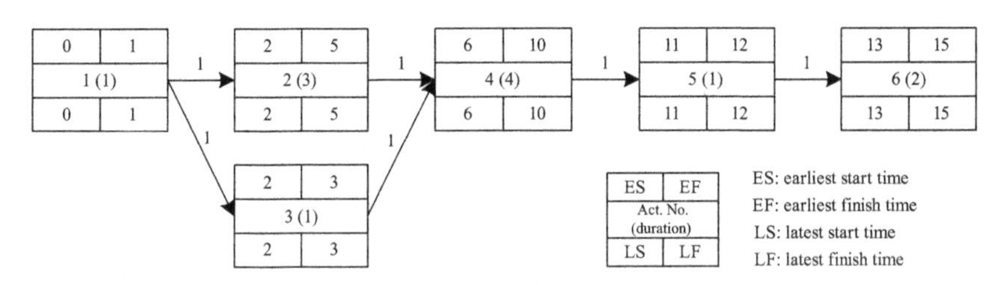

Figure 2.5 CPM calculations for the example application.

to be maintained between consecutive activities. The activities involved in the construction of one unit of the project are given, together with their estimated durations, in Table 2.1.

The example application is solved manually, applying the integrated CPM-LOB method, in the following steps:

- Perform CPM analysis for the first unit, considering unit duration of each activity and a minimum buffer time of one day. Figure 2.5 shows the CPM calculations for a single unit of the project, where the one-day buffer is set as a lag time between activities. The resulting CPM duration for the first unit (T_1) is 15 days and the critical path is 1-2-4-5-6. The total float values of noncritical activities are given in Table 2.2.
- Calculate the actual progress rate of each activity. Because the desired project duration (T_L) is 40 days, the desired progress rate of progress (R) can be calculated using Eq. (2.1) as $(10-1)/(40-15)=0.36$. The progress rate of noncritical activities is calculated considering total float using Eq. (2.2). The theoretical and actual number of crews, as well as actual progress rate of each activity, are calculated and are also given in Table 2.2.

Table 2.2 Line-of-Balance Calculations for the Example Application

No.	Duration D_i	Total Float TF_i	Actual Progress Rate $R_i = 9/(25 + TF_i)$	Theoretical Number of Crews $C_i = D_iR_i$	Actual Number of Crews C_{ai} = rounded up C_i	Actual Progress Rate $R_{ai} = C_{ai}/D_i$
1	1	0	0.36	0.36	1	1
2	3	0	0.36	1.08	2	0.667
3	1	2	0.333	0.333	1	1
4	4	0	0.36	1.44	2	0.5
5	1	0	0.36	0.36	1	1
6	2	0	0.36	0.72	1	0.5

- Draw the LOB schedule. Because there is no preceding activity for activity 1, it starts at time zero. The actual number of crews for this activity is 1, and its unit duration is 1 day. Thus, the last unit of activity 1 will be finished in the 10th day. The succeeding activities of activity 1 include activities 2 and 3, and their actual progress rates are not larger than 1. Then, the start times of both activities 2 and 3 in the first unit are equal to the summation of the finish time of activity 1 in the first unit and the buffer time of 2 days. Meanwhile, the finish times of activities 2 and 3 in the last unit are determined by their start times in the first unit and their actual numbers of crews, respectively. The resulting LOB diagram is shown in Figure 2.6. The start and finish times of each unit in each activity are given in Table 2.3.

2.4 COMMENTS AND FUTURE RESEARCH

The existing LOB technique is a heuristic procedure, since it cannot ensure that the deadline constraint is always satisfied. Take the project in the previous section as an example: the given deadline is 40 days, but the actual project duration obtained by the LOB calculation is 42 days. The common solutions for further shortening the project duration are: (1) increase or decrease the number of crews of some activities to improve or lower their progress rates; and (2) allow those activities with higher progress rates to be interrupted. In future studies, improvements of LOB techniques can be considered in the following ways:

- Integrate the time–cost trade-off analysis into LOB scheduling. When scheduling a project, planners always attempt to look for the

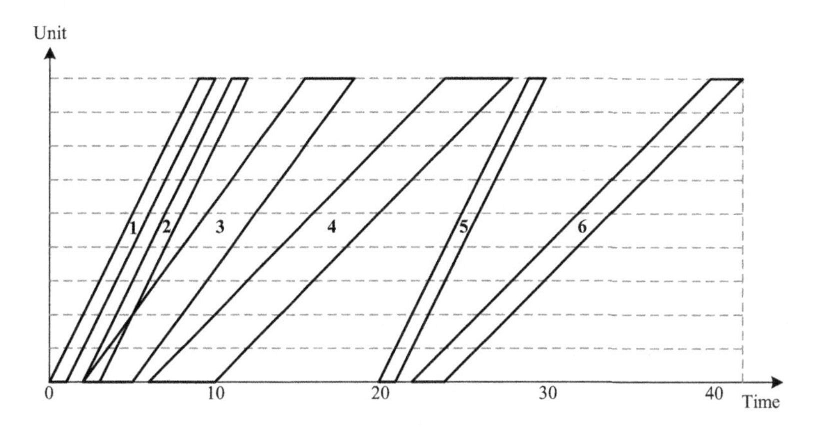

Figure 2.6 Line-of-balance diagram for the example application.

Table 2.3 Start and Finish Times for Each Sub-Activity

Activity	1		2		3		4		5		6	
Timing	ST	FT	ST	FT	ST	FT	ST	FT	ST	FT	ST	FT
1	0	1	2	5	2	3	6	10	20	21	22	24
2	1	2	3.5	6.5	3	4	8	12	21	22	24	26
3	2	3	5	8	4	5	10	14	22	23	26	28
4	3	4	6.5	9.5	5	6	12	16	23	24	28	30
5	4	5	8	11	6	7	14	18	24	25	30	32
6	5	6	9.5	12.5	7	8	16	20	25	26	32	34
7	6	7	11	14	8	9	18	22	26	27	34	36
8	7	8	12.5	15.5	9	10	20	24	27	28	36	38
9	8	9	14	17	10	11	22	26	28	29	38	40
10	9	10	15.5	18.5	11	12	24	28	29	30	40	42

optimal balance between time and cost for the project to be built. In LOB scheduling, the direct cost of an activity is proportional to the number of crews employed in this activity. The employment of more crews indeed results in a faster progress rate but obviously at an increased direct cost. On the other hand, violation of the resource continuity constraint by allowing work interruption may lead to an overall project duration extension but decrease the corresponding idle resource costs. In order to determine the optimum number of crews and interruption strategies for all activities so as to yield the minimum project cost while complying with a given deadline constraint, existing LOB techniques needs to be improved to have the ability to balance time and cost.

- Take the learning effect into consideration. Traditional LOB techniques assume that the production rate of an activity at each unit is constant. However, in many realistic applications, workers can improve their productivity with experience and practice (Lam et al., 2001; Jarkas, 2010; Pellegrino et al., 2012). As a result, the time and resources expended to complete the work on a unit will decrease as the number of repetitions increases. This phenomenon is known as "the learning effect" in the literature (Badiru, 1992). Considering the learning effect when planning and scheduling a project helps provide a realistic forecast of its duration and cost. This brings a higher degree of precision in budgeting and schedule, and can foster more competitive bidding. Thus, it is necessary to take the learning effect into consideration in LOB scheduling.
- Consider non-typical and nonrepetitive activities. Repetitive or linear construction, though it is characterized as a project of a repetitive nature, may contain some nonlinear and nonrepetitive activities. A non-typical activity is characterized by repetitive operations, where the output of operations is not uniform at every unit. For example, in a highway project, the workload of earthwork will vary from section to section, simply due to differences in the terrain. A nonrepetitive activity, on the other hand, is a one-off activity that does not repeat itself in every unit. An example of a nonrepetitive activity in a highway paving project is the posting of the occasional sign structure. Non-typical activities cannot be treated like typical and repetitive activities in LOB calculations because the outputs in these activities differ from unit to unit. The nonrepetitive portions of a project cannot be scheduled directly by the LOB method either, because these activities are not included in the CPM network of the first unit. Yet both non-typical and nonrepetitive activities may interfere with the scheduling of adjacent activities and, consequently, with the critical path. Therefore, the schedule for the entire project cannot be produced until these nonlinear and nonrepetitive activities are scheduled and coordinated with the typical and repetitive activities. There should therefore be a mechanism that allows the scheduler to accommodate non-typical and nonrepetitive activities in an LOB schedule (Arditi et al., 2002).

2.5 CONCLUSION

LOB is one of the most common tools for scheduling repetitive construction projects. The advantages lie in its ability to display progress rates and duration information for all activities in the LOB diagram. Executing LOB calculations aims to find a schedule that can satisfy the given deadline and resource continuity constraints for typical projects. Under some circumstances, the project duration obtained by the LOB calculation will be longer than the given deadline. At that point, the original schedule will need to be amended. Possible methods include (1) increasing (or decreasing) the number of crews of some activities, and (2) allowing some activities to be interrupted.

LOB is a scheduling tool waiting to be improved. The main limitations are: (1) it cannot perform a time–cost trade-off analysis; (2) it ignores the learning effect of workers by assuming the productivities of all crews remain unchanged; and (3) it cannot accommodate nontypical and nonrepetitive activities in an LOB schedule. These disadvantages greatly limit the application of LOB techniques in actual projects.

CHAPTER 3

Controlling Path Analysis in Repetitive Scheduling Method

3.1 INTRODUCTION

The controlling path, also known as the controlling activity path or controlling sequence, is defined as the longest continuous path from project start to project completion, which determines the minimum project duration under certain conditions (such as the crews that determine the production rate) and requirements (such as the constraints of precedence relations). The segments on the controlling path are named controlling segments, and the constraints (precedence relations) on the controlling path are named controlling constraints.

Harmelink and Rowings (1998) put forward a method for determining the controlling path of linear projects by upward pass and downward pass. The goal of the upward pass is to determine which activities or portions of activities could potentially be controlling. The process starts with the beginning of the project and progresses upward. In each step, the activity for which the potential controlling sub-activity is being determined is designated the origin activity, and the earliest point in time for this activity is designated as the origin. The next activity in the activity sequence list will be the target activity. The least distance interval describes the location at which this closest point occurs. Once the least distance interval has been determined, the point of intersection with the origin activity is called the critical vertex. The sub-activity of the origin activity between the origin and the critical vertex is then a potential controlling sub-activity for this activity, and the least distance interval becomes a potential controlling link between the origin and target activities. The target activity in this step of the upward pass becomes the origin activity for the next step, and the process repeats until all of the potential controlling activity segments have been determined. On the other hand, the purpose of the downward pass is to determine which portions of the potential controlling sub-activities are actually on the controlling path. This step can be

Repetitive Project Scheduling: Theory and Methods.

compared with the backward pass used in critical path method (CPM) scheduling, which identifies activities that do not have any float. In other words, in the case of linear activities on a linear schedule, the backward pass identifies sub-activities of activities for which the production rate cannot decrease without extending the duration of the project.

The scheduling method presented by Ammar and Elbeltagi (2001) considers both precedence relations and resource continuity constraints. The method utilizes the CPM network of a single unit, where start to start and finish to finish relationships are used. However, the method only applies to typical projects in which all units of an activity have the same work amount and the same duration.

Kallantzis and Lambropoulos (2004) developed a scheduling procedure for determining the controlling path in linear projects, where the maximum time and distance constraints are considered, in addition to the commonly used minimum time and distance constraints. The scheduling procedure includes four major steps. First, the procedure calculates the earliest finish day of the project, with the resource continuity constraints maintained and the specified production rates and constraints between activities ensured. Second, potential controlling activities are identified and their controlling sub-activities determined according to the relative positions of controlling points (CPs) with their successors and predecessors. Third, the maximum time and distance constraints are applied to the schedule. At this point, the activity with the highest production rate has to reduce its rate or introduce a certain number of interrupted days in order to comply with the maximum time and distance constraints. Finally, the controlling path is recomputed.

The above methods have been regarded as visual techniques lacking the analytical qualities of the CPM of scheduling (Harmelink and Rowings, 1998). To solve this problem, Lucko (2009) presented an integrated method of linear scheduling, called the "productivity scheduling method" (PSM). Considering all constraints, PSM uses singularity functions to mathematically describe activities and their buffers, and then automatically generates the overall project duration. Equation (3.1) provides the general model for modeling linear and repetitive activities and their buffers, which is shown in Figure 3.1.

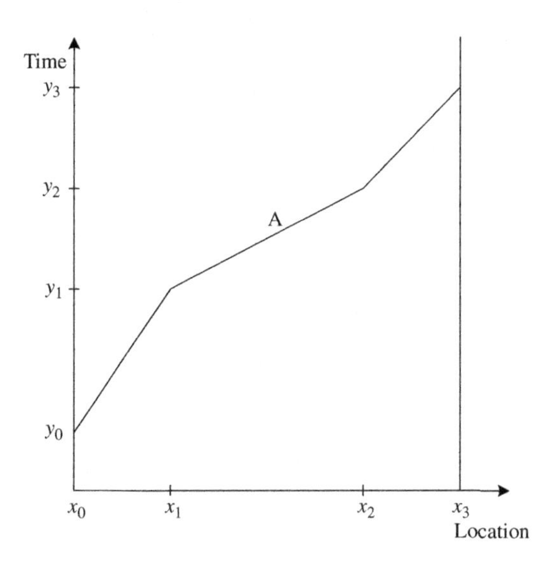

Figure 3.1 General model for singularity function.

$$
\begin{aligned}
y(x) &= y_0 \cdot \langle x-0 \rangle^0 + \frac{y_1 - y_0}{x_1 - x_0} \cdot \langle x-0 \rangle^1 \\
&\quad + \sum_{k=1}^{m-1} \left[\left(\frac{y_{k+1} - y_k}{x_{k+1} - x_k} - \frac{y_k - y_{k-1}}{x_k - x_{k-1}} \right) \cdot \langle x - x_k \rangle^1 \right] \\
&= y_0 \cdot \langle x-0 \rangle^0 + \frac{y_1 - y_0}{x_1 - x_0} \cdot \langle x-0 \rangle^1 + \left(\frac{y_2 - y_1}{x_2 - x_1} - \frac{y_1 - y_0}{x_1 - x_0} \right) \cdot \langle x - x_1 \rangle^1 \\
&\quad + \left(\frac{y_3 - y_2}{x_3 - x_2} - \frac{y_2 - y_1}{x_2 - x_1} \right) \cdot \langle x - x_2 \rangle^1 \\
&= \frac{1}{50} \cdot \langle x-0 \rangle^1 - \frac{3}{200} \cdot \langle x-100 \rangle^1 + \frac{1}{200} \cdot \langle x-300 \rangle^1
\end{aligned}
\tag{3.1}
$$

where x and y denote the amount variable and the time variable of an activity with m segments, respectively; y_k and x_k denote the pairs of coordinates with numbering index k. The summation term contains change terms, where the present slope y_k/x_k is replaced with a new slope y_{k+1}/x_{k+1}.

Singularity functions were originally used for structural engineering analysis of beams under complex loads. Equation (3.2) gives the basic term of singularity functions, written with angle brackets, as introduced by Wittrick (1965).

$$\langle x-a\rangle^n = \begin{cases} 0 & x<a \\ (x-a)^n & x\geq a \end{cases} \tag{3.2}$$

where a is the upper boundary of the current segment and exponent n denotes the order of phenomenon that changes at the end of the segment. The exponential rule $a^0 = 1$ applies to the brackets. Equations (3.3) and (3.4) describe how the brackets can be differentiated and integrated like regular mathematical functions.

$$\frac{\mathrm{d}}{\mathrm{d}x}\langle x-a\rangle^n = n\cdot\langle x-a\rangle^{n-1} \tag{3.3}$$

$$\int \langle x-a\rangle^n \,\mathrm{d}x = \frac{1}{n+1}\cdot\langle x-a\rangle^{n+1} + C \tag{3.4}$$

Although several methods have been proposed for determining the controlling path, there is still room to improve. First, different controlling sub-activities and controlling paths may be obtained for the same repetitive construction project using the methods mentioned above, which may confuse project planners and managers. Part of the problem may be caused by a distinct understanding of the controlling sub-activities and controlling path. Another reason for this problem can be due to errors in some methods in identifying the controlling path. In other words, the controlling path and controlling sub-activities identified do not conform to the time and distance constraints of the project. Finally, although it is known that the controlling sub-activities control the project duration, it is not clear how the controlling sub-activities control the project duration and how a change in a controlling sub-activity can change the project duration. In fact, different types of controlling sub-activities result in different consequences for project duration.

We propose a method to identify the controlling path and controlling sub-activities for repetitive construction projects using the repetitive scheduling method (RSM). The basis of this method is identifying potential CPs with constraints. Different types of controlling sub-activities and their properties are analyzed to investigate how the controlling sub-activities determine the project duration and how a change in a controlling sub-activity changes project duration.

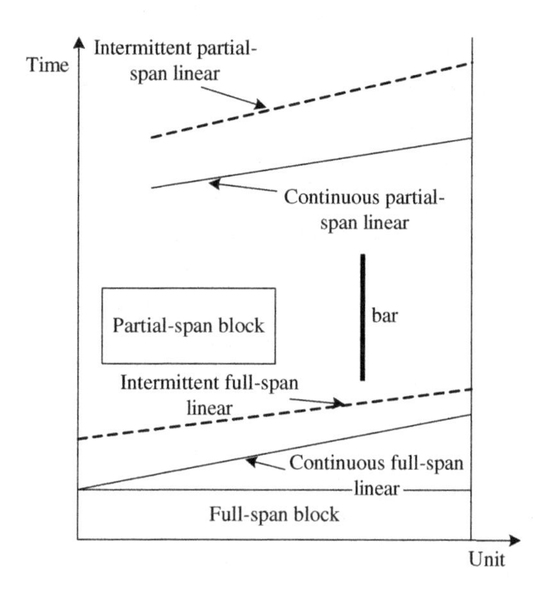

Figure 3.2 Activity types in RSM.

3.2 BASIC REPRESENTATION OF RSM

RSM is a two-dimensional coordinate system, in which the horizontal and vertical axes represent production unit and time, respectively. There are three types of activities that can appear in a linear or repetitive schedule: linear, block, and bar (Vorster et al., 1992). Harmelink and Rowings (1998) refined the linear activity type into four specific subtypes: continuous full-span linear, intermittent full-span linear, continuous partial-span linear, and intermittent partial-span linear. Block-type activities is divided into two types: full-span block and partial-span block.

These subtypes relate to whether or not an activity spans the entire location of the project and whether or not the activity is in continuous or intermittent operation. Figure 3.2 shows all of the activity types that would appear on a linear or repetitive schedule.

3.3 METHOD FOR DETERMINING CONTROLLING PATH

Consider the following assumptions that are also adopted by existing studies:

1. The repetitive construction project to be analyzed includes I activities labeled $i = 1,2,\ldots,I$, and J units labeled $j = 1,2,\ldots,J$; the activity types are linear, block, and bar.

2. Minimum time constraints between activities contain four types: SS, SF, FS, and FF. These constraints can be described mathematically by the following equations, where s_{ij} and f_{ij} denote the start time and the finish time of activity i in unit j; T_{ti} denotes the lag time between activity i and its preceding activity t.

$$\text{(SS)} \quad s_{tj} + T_{ti} \le s_{ij}, \quad j = 1, 2, \ldots, J \tag{3.5}$$

$$\text{(SF)} \quad s_{tj} + T_{ti} \le f_{ij}, \quad j = 1, 2, \ldots, J \tag{3.6}$$

$$\text{(FS)} \quad f_{tj} + T_{ti} \le s_{ij}, \quad j = 1, 2, \ldots, J \tag{3.7}$$

$$\text{(FF)} \quad f_{tj} + T_{ti} \le f_{ij}, \quad j = 1, 2, \ldots, J \tag{3.8}$$

3. Minimum distance constraints are described as two activities that cannot approach each other more than a specified amount of unit length at any time during the project duration, which can also be described mathematically by Eq. (3.9), in which D_{ti} denotes the amount of distance constraint between activity i and its preceding activity t.

$$s_{t,j+D_{ti}} \le s_{ij}, \quad j = 1, 2, \ldots, J - D_{ti} \tag{3.9}$$

$$f_{t,j+D_{ti}} \le f_{ij}, \quad j = 1, 2, \ldots, J - D_{ti} \tag{3.10}$$

4. Maximum (time and distance) constraints are not considered.
5. All activities must satisfy the resource continuity constraint.
6. The work sequence for all activities is from unit 1 to unit J.
7. Only one crew is employed for each activity.

The CP is the basis for identifying the controlling path. A CP is defined as the event on a controlling activity linking another controlling activity. Usually there are two CPs on a controlling activity. One is the CP linking its preceding controlling activity, called the preceding CP. The other is the CP linking its succeeding controlling activity, called the succeeding CP. Specifically, for the first controlling activity, the starting point is defined as its preceding CP. For the last controlling activity, the finishing point is defined as its succeeding CP.

For a project with all the production rates of its activities known, its duration is determined by the constraints between the controlling activities. So a CP must be the point where the constraint takes effect. On the other hand, the points on an activity where the constraints take effect may be identified to be CPs, depending on whether the activity is a controlling activity. Therefore, these CPs are called the potential CPs (pCPs). The process of determining pCP is presented below.

3.3.1 Determining pCPs with Time Constraint

In Figure 3.3(a), activity i and its preceding activity t satisfy the time constraint of FS with a lag time of T_{ti}. So the starting points of activity i at every unit are constrained by the relative finishing points of activity t. By shifting all the finishing points of activity t to the left for a unit, we could match the finishing points of t and the starting points of i. Then moving all the points upward for T_{ti} days, and linking the points, a dashed line is obtained, shown as the dashed line in Figure 3.3(a). The dashed line is defined as the time constraint line of activity t to i, denoted by TCL($t \rightarrow i$). Therefore, the starting points of i are forbidden to enter the shaded area between TCL($t \rightarrow i$) and activity t.

Move activity i downward until the first touch point with constraint line TCL($t \rightarrow i$). The first touch point of activity i is the pCP on activity i from activity t, denoted by pCP($i \leftarrow t$). Similarly, the corresponding finishing point in the same unit on activity t is also identified as the pCP, denoted by pCP($t \rightarrow i$).

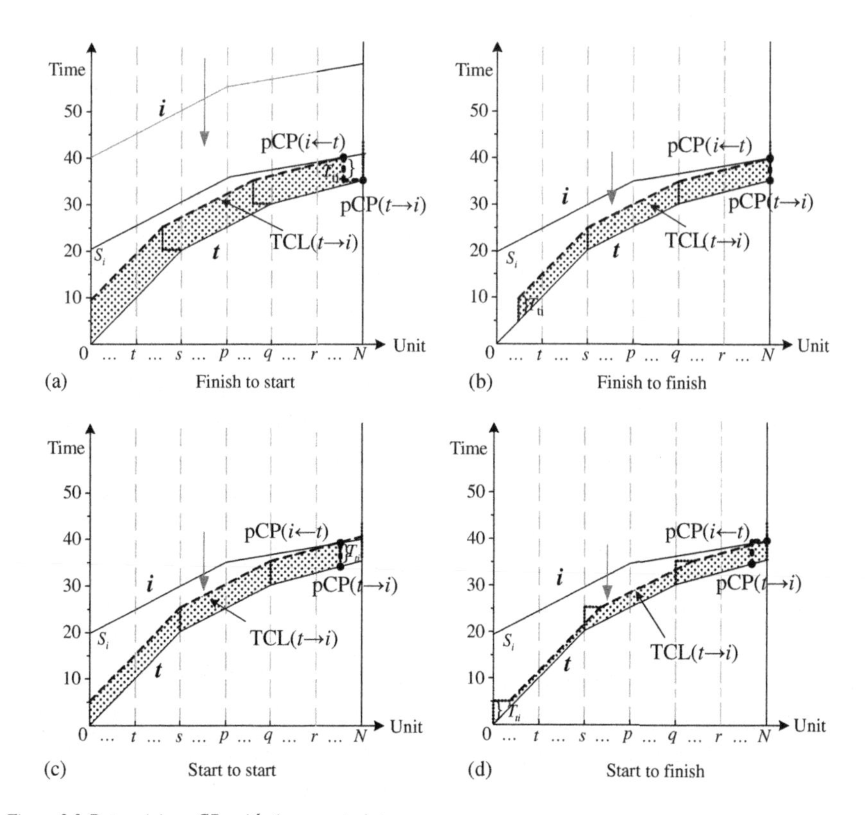

Figure 3.3 Determining pCPs with time constraint.

After the identification of pCP($i \leftarrow t$), the earliest start time of activity i is identified as well. Since the earliest start time cannot be negative, it should be set as zero, if less than zero. As a result, there is no touch point and the constraint line between the activities, and no pCP either. When the earliest start time of activity i is determined, its earliest finish time can be determined as well, with the production rate known. Therefore, the process of determining pCPs is also the process of determining the earliest start time and finish time of an activity. When a pair of pCPs are identified as CPs, the relationship that represents the constraint and connects the pCPs together in a graph is defined as the controlling constraint, shown as bold dashed lines in Figure 3.3(a).

If the time constraints between activity t and i are FF, SS, or SF relations, pCPs can be obtained in a similar way, as shown in Figures 3.3(b)–(d), respectively. Though the time constraint amount is the same, different constraint types result in different pCPs and different earliest start times for activity i. Consequently, it is crucial to identify the constraint type between activities to identify pCPs and determine the controlling path.

The method above to identify pCPs belongs to a graphic method, and cannot handle large-scale projects effectively. Thus, an equivalent mathematical algorithm is proposed. For activity t and its succeeding activity i, the algorithm aims at determining the coordinates of pCP($i \rightarrow t$) and pCP($t \leftarrow i$) as well as the start time s_{ij} of activity i in each unit j, automatically. Suppose that the set D_i ($i = 1, \ldots, I$) consists of the labels of those units in which activity i has a working task; d_{ij} represents the duration of activity i in unit j. The algorithm is shown as follows:

Algorithm 3.1 Calculating the pCPs Meeting the Time Constraints

1. Initialize parameter λ and γ. If activity t and its succeeding activity i are constrained by time constraints of SF or FF, $\lambda = 1$; otherwise, $\lambda = 0$. If activity t and i are constrained by time constraints of FS or FF, $\gamma = 1$; otherwise, $\gamma = 0$.
2. Initialize the start time of activity i in each unit j by considering the resource continuity constraint.

$$s_{ij} = \sum\nolimits_{k \in \Omega_j} d_{ik}, \quad \Omega_j = \{j^* \mid j^* < j, j \in D_i\}$$

$$\left(\text{presuming } \sum\nolimits_{k \in \varnothing} d_{ik} := 0\right);$$

3. Calculate the minimum amount of time, denoted by Δ_i (the start and finish times of activity i in each unit should be delayed in order to meet the time constraints), and determine the positions of pCPs, denoted by j^p.

$$\Delta_i = \max_{j \in D_i \cap D_t} \{s_{tj} + \gamma d_{tj} + T_{ti} - s_{ij} - \lambda d_{ij}, 0\},$$

$$j^p = \{j | s_{tj} + \gamma d_{tj} + T_{ti} - s_{ij} - \lambda d_{ij} = \Delta_i,\ j \in D_i \cap D_t\}.$$

4. Revise the start and finish times of activity i in each unit j according to Δ_i. Assume that the first working unit of activity i is j'; then $s_{ij} := s_{ij} + \Delta_i$ and $f_{ij} = s_{ij} + d_{ij}$ for all $j = j', \ldots, J$.
5. Identify the coordinates of pCPs:

$$\text{pCP}(t \to i) = (j^p - 1 + \gamma,\ s_{tj^p} + \gamma d_{tj^p})$$
$$\text{pCP}(i \leftarrow t) = (j^p - 1 + \lambda,\ s_{ij^p} + \lambda d_{ij^p})$$

3.3.2 Determining pCPs with Distance Constraint

Figure 3.4(a) shows a distance constraint between activity t and its succeeding activity i with the constraint amount D_{ti}. The distance constraint line between t and i can be obtained in the similar way as the time constraint. Shift all the points of activity t to the left for a distance D_{ti}, as the dashed line shown. The dash line is defined as the distance constraint line of activity t to activity i, denoted by DCL($t \to i$). Both the starting and finishing points of activity i are forbidden to enter the shaded area between DCL($t \to i$) and activity t.

Shift activity i downward to a touch point with the constraint line DCL($t \to i$). The touch point of activity i is the pCP on activity i from activity t, denoted by pCP($i \leftarrow t$). Similarly, the corresponding constraint point on activity t in the same unit is identified as a pCP, denoted by pCP($t \to i$).

For the distance constraint between activity t and its succeeding activity i, with coordinates of pCP($i \to t$) and pCP($t \leftarrow i$), the start time of activity i in each unit can be determined by the following mathematical algorithm. Note that set D_i ($i = 1,2,\ldots,I$) consists of the labels of those units in which activity i has a working task.

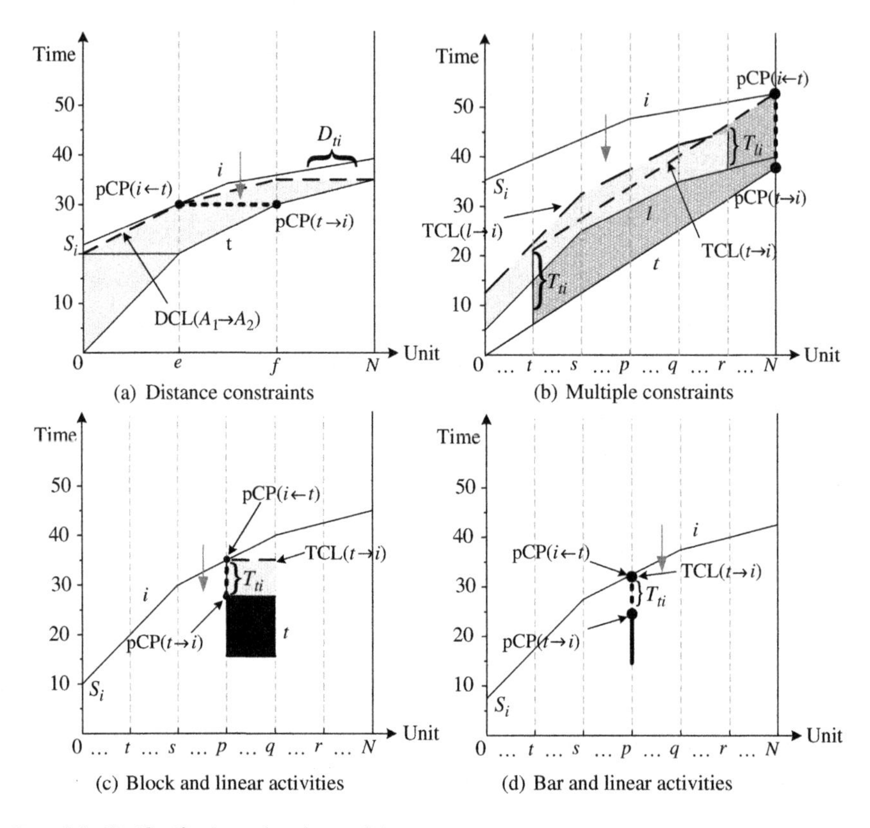

Figure 3.4 pCPs identification under other conditions.

Algorithm 3.2 Calculating pCPs Satisfying Distance Constraint

1. Determine the range of the distance constraint line DCL($t \rightarrow i$).

$$\overline{D} = \{j | j + D_{ti} \in D_t, j > 0\};$$

2. Initialize the start time of activity i in each unit j by considering the resource continuity constraint. This step is the same as that in Algorithm 3.1.
3. Calculate the minimum amount of time (Δ_i) that the start and finish times of activity i in each unit should be delayed in order to meet the distance constraint, and determine the positions of pCPs, j^p.

$$\Delta_i = \max_{j \in D_i \cap \overline{D}} \{s_{t,j+D_{ti}} - s_{ij},\ s_{t,j+D_{ti}} + d_{t,j+D_{ti}} - s_{ij} - d_{ij}, 0\},$$

$$j_1^p = \{j | s_{t,j+D_{ti}} - s_{ij} = \Delta_i,\ j \in D_i \cap \overline{D}\},$$

$$j_2^p = \{j | s_{t,j+D_{ti}} + d_{t,j+D_{ti}} - s_{ij} - d_{ij} = \Delta_i,\ j \in D_i \cap \overline{D}\},$$

$$j^p = j_1^p \cup j_2^p;$$

4. Revise the start and finish times of activity i in each unit j according to Δ_i. This step is the same as that in Algorithm 3.1.
5. Identify the coordinates of pCPs:

$$\text{If } j_1^p \neq \emptyset,\ \text{pCP}(t \rightarrow i) = \left(j_1^p - 1 + D_{ti},\ s_{t,j_1^p - 1 + D_{ti}}\right),$$
$$\text{pCP}(i \leftarrow t) = \left(j_1^p - 1,\ s_{i,j_1^p - 1}\right);$$

$$\text{If } j_2^p \neq \emptyset,\ \text{pCP}(t \rightarrow i) = \left(j_2^p + D_{ti},\ s_{t,j_2^p + D_{ti}} + d_{t,j_2^p + D_{ti}}\right),$$
$$\text{pCP}(i \leftarrow t) = \left(j_2^p,\ s_{ij_2^p} + d_{ij_2^p}\right).$$

3.3.3 Determining pCPs with Multiple Constraints

If an activity is constrained by multiple constraints, it is necessary to find out the active constraint. The active constraintwhich could can be obtained done by in the following way. Find out all the time and distance constraint lines first, and then shift the constrained activity downward until it touching touches with any constraint line. In Figure 3.4(b), activity i has a time constraint of FF with its preceding activity t and a time constraint of SS with its preceding activity l. The time constraint lines of TCL($t \rightarrow i$) and TCL($l \rightarrow i$) are represented by different types of dashed lines. Move activity i downward, and it touches TCL($t \rightarrow i$) first. Thus, the constraint between activities t and i is the active constraint determining the earliest start time of activity i and the position of pCPs.

3.3.4 Determining pCPs with Constraint with Bar Activity and Block Activity

If constraints exist between linear activity and block activity or between linear activity and bar activity, the pCPs can be identified in a similar way. The constraint line of a block activity is parallel to the bottom edge of the activity, while the constraint line of a bar activity is only a point, as shown in Figures 3.4(c) and (d).

3.3.5 Identifying the Controlling Path

For a repetitive construction project with all the production rates of the activities and the constraints between activities known, the controlling path can be identified by the following steps, where Steps 1–3 determine the start times of activities in all units and pCPs, while Steps 4–6 determine CPs and controlling path by back tracing.

Step 1. Identify the durations of all activities in each unit, as well as constraint types and values between activities.
Step 2. If activity i has no predecessor, start at time zero, and determine its starting point as the pCP; otherwise, for every predecessor of activity i, $t_k(k=1,\ldots,K)$, the minimum amount of time activity i should be delayed, denoted by Δ_i^k, is calculated by Algorithm 3.1 or 3.2. If $\Delta_i^{k'} = \max_{k=1,\ldots,K}\left(\Delta_i^k\right)$, the constraint between activity k' and i its controlling constraint, pCPs and the start times of activity i in every unit are then determined.
Step 3. If the time parameters of all activities are calculated, then initialize set $\beta=\varnothing$ and go to the next step, or go back to Step 2.
Step 4. To find the activity with the longest finishing time, determine its finish point as a CP. If this activity has predecessors, put all the corresponding pCPs located on its predecessors into set β and go the Step 5; otherwise, identify the start point of this activity as a CP and go to Step 6.
Step 5. If $\beta=\varnothing$, move to the next step; otherwise, for any pCP($i \leftarrow t$)$\in\beta$, confirm pCP($i \leftarrow t$)$\in\beta$ and pCP($t \rightarrow i$) as CPs, and identify the constraint line connecting the two CPs as a controlling constraint. Then remove pCP($i \leftarrow t$) from set β. If activity t has a predecessor, put all the corresponding pCPs located on its predecessors into set β, and then continue; otherwise, identify the start point of activity t as a CP and go to the next step.
Step 6. The sub-activity between two CPs on the same activity is identified as the controlling sub-activity. Then the controlling path can be obtained by linking all the controlling sub-activities and constraints.

3.4 TYPES OF SUB-ACTIVITIES

After the controlling path is determined, controlling sub-activities can be divided into three types:

1. *Forward controlling sub-activity*: If the preceding CP is realized earlier than the succeeding CP on a controlling activity, the controlling

sub-activity between these two CPs is a forward controlling sub-activity. For a controlling activity, the forward controlling sub-activity means its preceding CP lies below its succeeding CP. The forward controlling sub-activity is similar to the critical activity in the critical path method network. The project duration will change in the same direction as that of the forward controlling sub-activity; that is, if a forward controlling sub-activity is prolonged, the project duration will be prolonged.

2. *Point controlling sub-activity*: If the preceding CP coincides with the succeeding CP on a controlling activity, the CP is defined as a point controlling sub-activity. For the controlling sub-activity, only the time when the CP is realized affects the project duration. If it is delayed, the project will be delayed.
3. *Backward controlling sub-activity*: If the preceding CP is realized later than the succeeding CP on a controlling activity, the controlling sub-activity between these two CPs is a backward controlling sub-activity. For a controlling activity, the backward controlling sub-activity means its preceding CP lies above its succeeding CP. The backward controlling sub-activity has a special property. At the planning stage, variation of the duration of a backward controlling sub-activity will change the project duration in the opposite direction; that is, if the duration of a backward controlling sub-activity is prolonged, the project duration could be reduced. This is because when the duration of the backward controlling sub-activity is prolonged, the succeeding CP could be realized earlier without violating the constraint from the preceding activity. Thus the succeeding activity could be started earlier and the project could be finished earlier. In network modeling this kind of activity is defined as backward critical activity by Elmaghraby and Kamburowski (1992).

3.5 PROJECT DURATION DETERMINATION

When the controlling path is identified, the project duration, denoted by D, can be represented by

$$\begin{aligned} D = & \sum \text{durations of all forward controlling sub-activities} \\ & - \sum \text{durations of all backward controlling sub-activities} \\ & + \sum \text{lag times of all controlling time constraints} \end{aligned} \tag{3.11}$$

3.6 CASE STUDY

There is a highway construction project, with the highway extending 1500 m. It contains nine activities: ditch excavation, culvert, concrete removal, peat excavation and swamp backfill, embankment, utility work, subbase, gravel, and paving. If the length of one unit is 60 m, the project includes 25 units. Specific project information is shown in Table 3.1.

As illustrated in Table 3.1, bar activity 2 (i.e., culvert) has no preceding activity, so it begins at time zero. There is a time constraint of FS with lag time of 1 day between linear activity 1 and bar activity 2. The time constraint line, TCL($2 \rightarrow 1$), of bar activity 2 to linear activity 1 is just a point. Shift activity 1 downward until it touches the constraint line TCL($2 \rightarrow 1$), so the start time of activity 1 in the first unit is -1 d, which should be reset as 0. Consequently, there is no pCP between activities 1 and 2. Repeat this step through the last activity. The earliest start of each activity in the first unit, the earliest finish time of each activity in the last unit, and the coordinates of pCPs are shown in Table 3.2.

Table 3.1 Project Information

Activity	Name	Workspace (m); Work Efficiency (m/d)	Constraint	Type	Position
1	Ditch excavation	0–720 (m): 360 (m/d)	FS_{21} = 1d	Linear	Whole section
		720–1500: 260			
2	Culvert	2 d		Bar	1260 m
3	Concrete removal	1–1500: 150	SS_{13} = 2d	Linear	Whole section
4	Peat excavation and swamp backfill	6 d	FS_{34} = 2d	Block	240–360 m
5	Embankment	1–600: 100	FS_{45} = 2d	Linear	Whole section
		600–1500: 225			
6	Utility work	900–1500: 300	FF_{56} = 1d	Linear	900–1500 m
7	Subbase	0–900: 113	FS_{67} = 1d	Linear	Whole section
		900–1500: 300	DC_{57} = 60m		
8	Gravel	1–1500: 313	FF_{78} = 2d	Linear	Whole section
9	Paving	1–1500: 250	FS_{89} = 1	Linear	Whole section

Table 3.2 The Earliest Time Parameters and pCPs of Each Activity

Activity	Earliest Start Time in the First Unit (d)	Earliest Finish Time in the Last Unit (d)	pCPs Corresponding to Predecessors	pCPs Corresponding to Successors
1	0	5	—	pCP$(1\rightarrow3)(0,0)$
2	0	2	—	—
3	2	12	pCP$(3\leftarrow1)(0,2)$	pCP$(3\rightarrow4)(360,4.4)$
4	6.4	12.4	pCP$(4\leftarrow3)(360,6.4)$	pCP$(4\rightarrow5)(240,12.4)$
5	12	22	pCP$(5\leftarrow4)(240,14.4)$	pCP$(5\rightarrow6)(1500,22)$
				pCP$(5\rightarrow7)(600,18)$
6	21	23	pCP$(6\leftarrow5)(1500,23)$	—
7	15.4	25.3	pCP$(7\leftarrow5)(300,18)$	pCP$(7\rightarrow8)(1500,25.3)$
8	22.5	27.3	pCP$(8\leftarrow7)(1500,27.3)$	pCP$(8\rightarrow9)(60,22.7)$
9	23.7	29.7	pCP$(9\leftarrow8)(0,23.7)$	—

Determine CPs and the controlling path by back tracing. First, determine the finish point of activity 9 as a CP. Find the only pCP on activity A_9, resulting from the time constraint between activities 8 and 9. Thus, identify pCP$(9\leftarrow8)$ and pCP$(8\rightarrow9)$ as CPs, denoted by CP$(9\leftarrow8)$ and CP$(8\rightarrow9)$. Repeat this step for the first activity. Finally, connect all controlling sub-activities and constraints to obtain the controlling path, shown as the bold line in Figure 3.5. The project duration is 29.7 d.

Comparing Figure 3.3 with Figure 3.4, the finishing point of activity 6 in the last unit is identified as the pCP, but it is not determined as the CP. Thus, activity 6 contains no controlling sub-activity. The controlling sub-activity on activity 1 is a point controlling sub-activity, since its preceding CP coincides with the succeeding CP. Because the realization time of the preceding CP of activity 8 is later than the succeeding CP, the controlling sub-activities on activity 8 are backward controlling sub-activities. Obviously, controlling sub-activities on other activities are forward controlling sub-activities.

3.7 DISCUSSION

The method for identifying the controlling path proposed in this paper is a development on the basis of the methods proposed by Harris and Ioannou (1998), Harmelink and Rowings (1998), and Lucko (2009). Compared with those methods, it has the following advantages.

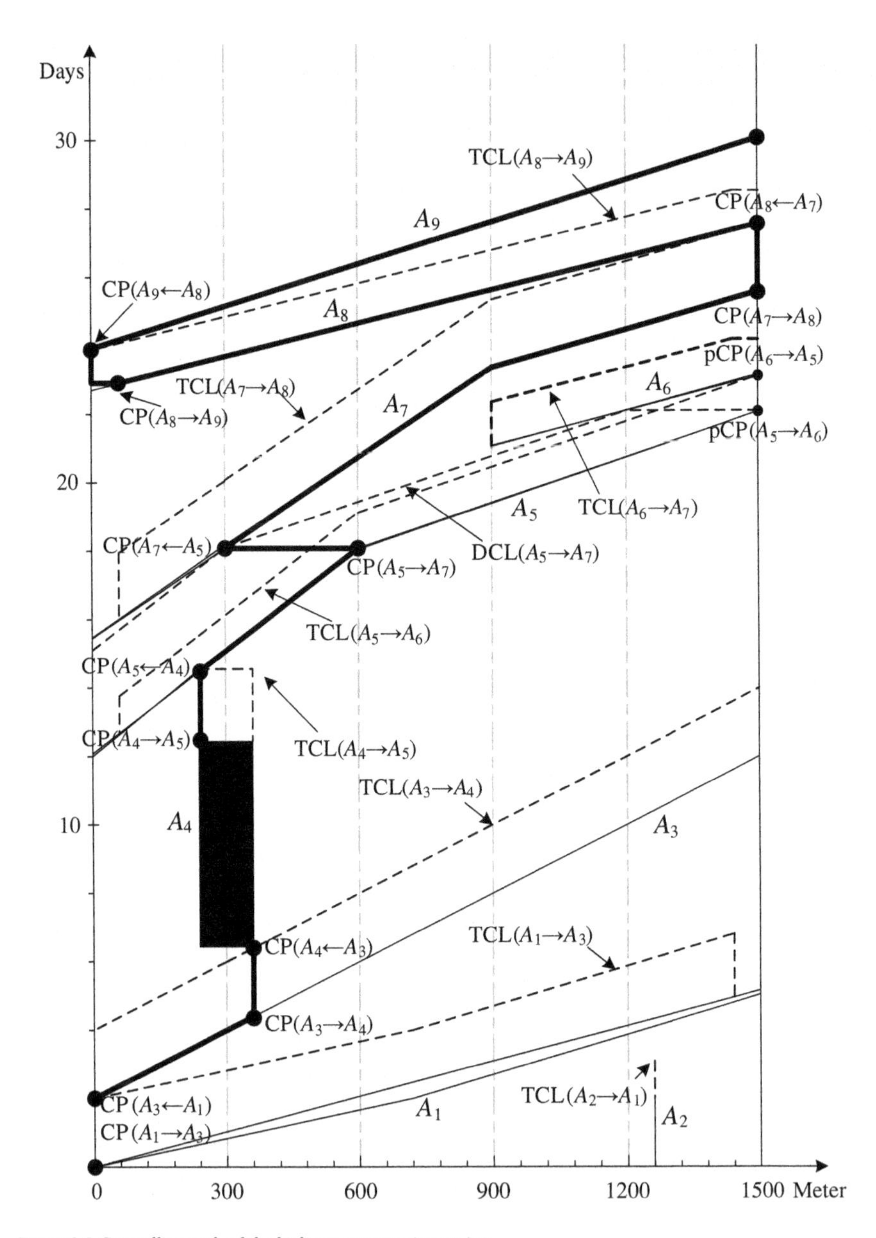

Figure 3.5 Controlling path of the highway construction project.

First, it presents a correct way to identify the controlling path and controlling sub-activities conforming to the requirements of the project. There are some problems in the method proposed by Harmelink and Rowings (1998). Sometimes it incorrectly identified pCPs and controlling constraints, which would lead to incorrect identification of the

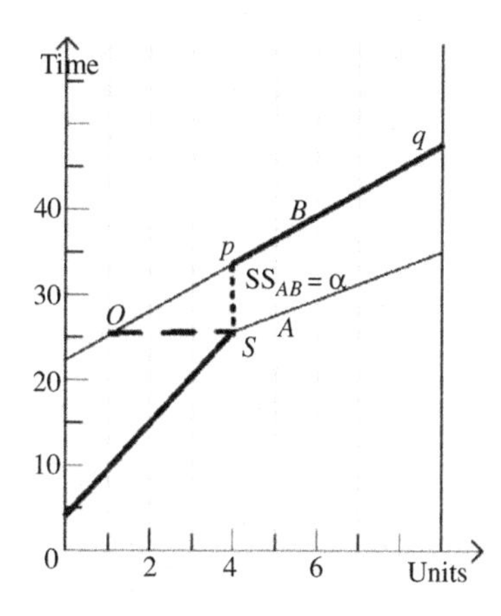

Figure 3.6 Different CPs and different controlling paths.

controlling path and controlling sub-activities. For example, assume there are only two activities *A* and *B* in a project, and the minimum time constraint between them is start to start with α lag time, as shown in Figure 3.6. According to the method of Harmelink and Rowings (1998), point *o* instead of point *p* is identified as the pCP($B \leftarrow A$). So sub-activity *oq* is identified as the controlling sub-activity and the dashed line *os* is identified as the controlling constraint between activities *A* and *B*. In fact, the line *os* represents the distance activity *A* leads activity *B* at that time. So only the line *ps* could correctly represent the controlling constraint. The sub-activity *op* does not belong to the controlling sub-activity.

Furthermore, Harmelink and Rowings (1998) assume that the minimum distance interval always intersects the minimum time interval. However, when production rates change, the minimum time interval and minimum distance interval often occur at different points and they do not intersect. For example, in Figure 3.7, the minimum distance interval between activities *A* and *B* appears at point *o* on activity *A*, but the minimum time interval lies at point *q* on activity *A*. Which point would be the CP is determined by the constraint types between activities *A* and *B*. As this method converts all the time constraints to distance constraints, the controlling path and controlling sub-activities identified by this method do not conform to the project itself.

In some repetitive construction projects, there are explicit precedence relations (or constraints) between the activities. Therefore, it is

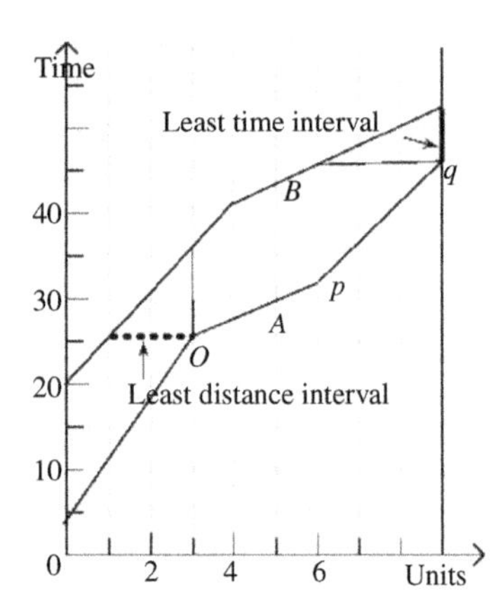

Figure 3.7 Least time interval and least distance interval.

necessary to distinguish the different types of constraints while identifying the controlling path. Lucko (2009) used a buffer running through all the units, which may actually be stricter than any kind of time constraint. It is in fact a constraint of SS ∪ FF, which may lead to a later finishing time than the project actually needs. For example, the constraint between activities A and B is start to start with α lag time; the pCPs and the starting time of activity B are determined as shown in Figure 3.8. The start time of activity B is later according to Lucko's method, and the pCPs are changed as well, as shown in Figure 3.9.

Second, identifying the pCPs by constraint lines offers more convenience. Some methods assume that repetitive activities seldom change their production rates. As a result, the pCPs are identified by observing whether the consecutive activities are converging or diverging (Harris and Ioannou, 1998; Harmelink and Rowings, 1998). In fact, the production rate of an activity may change from unit to unit because of different amounts of work content in each unit, weather, change of crew size, the learning effect, and other factors. It is difficult to tell whether the consecutive activities are converging or diverging. Using constraint lines, it is convenient to deal with all kinds of production rates.

Finally, this method provides a way to identify the three types of controlling sub-activities and disclose how the controlling sub-activities determine project duration. Most existing methods have not paid enough attention to the backward and point controlling

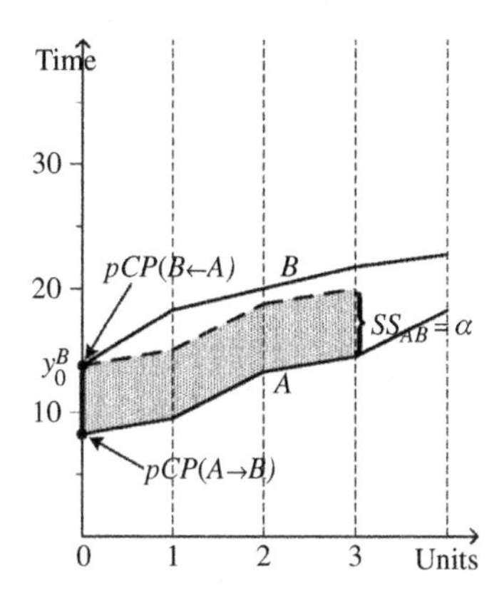

Figure 3.8 The pCPs determined by the current method.

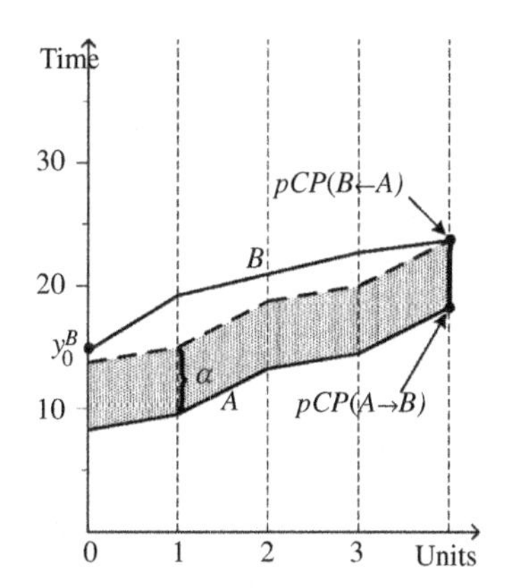

Figure 3.9 The pCPs determined by Lucko's method.

sub-activities; Hamelink and Rowings (1998) even rejected the backward controlling sub-activity from the controlling path during the process of the downward pass.

3.8 CONCLUSION AND PROSPECTS

The controlling path and controlling sub-activities are the basis of the repetitive construction project scheduling. This chapter has proposed a method for identifying the controlling path based on the technology of identifying the pCPs, and a comparison with the existing methods has been made. It has also disclosed how the project duration is determined through analysis of the three types of controlling sub-activities, namely the forward controlling sub-activity, the backward controlling sub-activity, and the point controlling sub-activity. The proposed method is suitable for both linear projects and vertical repetitive construction projects.

People usually focus on the forward controlling sub-activities in project scheduling. The properties of the backward controlling sub-activities and point controlling sub-activities can be used in the minimum project duration problem. These are topics for future research.

CHAPTER 4

Conversion of Repetitive Scheduling Model to Network Model

4.1 INTRODUCTION

The repetitive scheduling model (RSM) provides a way of scheduling repetitive construction projects, as an alternative to the commonly used network models. Although RSM is more visual, straightforward, and easier to use, network models are widely accepted, being used by both owners and construction contractors, and are often required as part of the construction contract in the field of project management. Therefore, it is important for practitioners to understand the function of the two methods in this area. Moreover, if RSM can be transformed to an equivalent network model, practitioners can take advantage of both methodologies.

Work has been done in converting RSM to the network model by Ammar and Elbeltagi (2001), Gransberg (2007), and Kallantzis et al. (2007). However, to the best of our knowledge, there is currently no complete method for successfully transferring an RSM to an equivalent network model. In making such a conversion, it is most important that the controlling path of the RSM coincide with the critical path of the network model.

Yamin and Harmelink (2001) presented a comparison between RSM and the critical path method (CPM) in such aspects as ease of use, accuracy in calculations, and critical paths. Two small examples are used to compare the controlling path in RSM and the critical path in network model. However, only one of them is a three-activity CPM network transformed into an equivalent linear project. Ammar and Elbeltagi (2001) constructed a precedence network, equivalent to the repetitive diagram, by designating activities finish to finish (FF), start to start (SS), or both FF and SS precedence relations depending on the production rates of their predecessor(s) and successor(s). The proposed methodology was applied on a sample project, but variable production rates within the same activity were not allowed. Gransberg (2007)

Repetitive Project Scheduling: Theory and Methods.

proposed a process to convert the repetitive schedules to critical path methods with precedence relations, but no comparison was made between the controlling path in RSM and the critical path in CPM. Kallantzis et al. (2007) presented a controlling/critical path comparison between the Kallantzis–Lambropoulos repetitive project model and CPM. Instead of using one or two illustrative examples and comparing the controlling/critical paths, a group of 25 multi-rate random linear projects was examined. Results showed that the equivalent repetitive projects produced different controlling paths and longer durations compared to their CPM networks. However, when the resource continuity constraint was removed, project execution times and controlling/critical paths coincided.

Differences between RSM and CPM also exist in terms of activity criticality. In CPM, a project will be delayed if the critical activity is delayed, while, as pointed out by Harris and Ioannou (1998) and Kallantzis et al. (2007), the makespan of an RSM project can paradoxically be shortened if the durations of some controlling activities are increased. These activities are the backward controlling segments defined in Chapter 3.

In this chapter, a method for transforming an RSM into the equivalent network model is developed, following which a critical path comparison is made between the RSM and the network model for three cases, each with differing resource continuity requirements. The results of this comparison show that the RSM completely coincides with the network model. Finally, the cause of the differences seen in the relevant literature between the RSM and the network model is determined.

4.2 METHOD FOR CONVERTING RSM TO NETWORK MODEL

An RSM usually contains two types of relations: (1) logical relations between units performing the same activity (i.e., the logical sequence from one unit to another given existing resource continuity constraints), and (2) precedence relations, which regulate the constraints—including the distance and time constraints—between different activities. The method proposed here needs to convert all activities and relations in an RSM into those of a network model. In this chapter, the network under generalized precedence relations (GPR) (Elmaghraby and Kamburowski, 1992) is adopted.

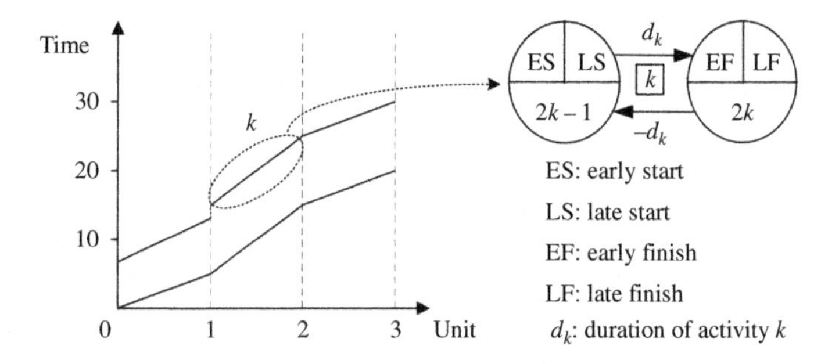

Figure 4.1 Sub-activity conversion.

4.2.1 Conversion of Activities

As shown in Figure 4.1, each activity in the RSM is separated into several sub-activities, each of which corresponds both to a unit of the RSM and to an activity in the GPR network. In the GPR network, each activity has two solid arcs in opposite directions. The forward arc, carrying a positive value, represents the minimum duration of the activity, and the backward arc, carrying a negative value, represents the maximum duration. In this paper, it is assumed that each sub-activity k, shown in Figure 4.1, has a fixed duration d_k. Therefore, two arcs in the GPR network carry the same absolute value d_k with one positive and one negative.

4.2.2 Conversion of Logical Relations

The logical sequence of an activity in RSM determines the order in which sub-activities are performed. Without loss of generality, the logical sequence from unit 1 to unit J is adopted for all activities, where J represents the total number of a repetitive construction project. At this time, the sub-activity of each activity in unit $j + 1$ cannot start until the completion of this activity in unit j, and this sequence can be converted into the minimum time constraint of FS with zero lag time in the GPR network.

In some cases, an activity in the RSM is required to maintain resource continuity; that is, the succeeding sub-activity in unit $j + 1$ must start immediately after the preceding sub-activity in unit j has finished. As (to the best of our knowledge) no existing method represents the resource continuity, in this study, the maximum time constraint of FS with zero lead time in the GPR network is used to represent this continuity.

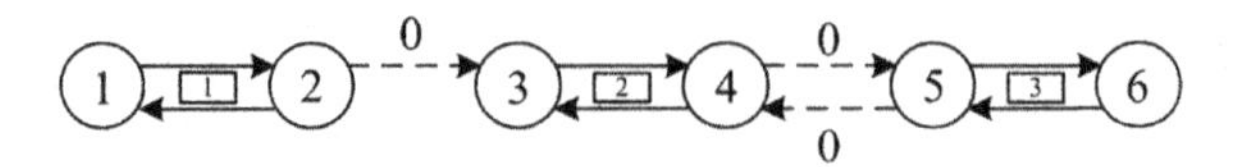

Figure 4.2 Conversion of logical relations.

In the GPR network, a forward arc with a non-negative value and a backward arc with a non-positive value are used to represent the minimum and maximum time constraints, respectively. As shown in Figure 4.2, three sub-activities of activity A are A_1, A_2, and A_3. There is no resource continuity constraint between sub-activities A_1 and A_2, but there is a resource continuity constraint between A_2 and A_3. Then the logical sequence is represented by using a forward arc connecting node 2 and node 3 and a forward arc connecting node 4 and node 5. The resource continuity constraint between sub-activities A_2 and A_3 is represented by applying a backward arc connecting node 4 and node 5.

4.2.3 Conversion of Precedence Relations

Different from the GPR network, the precedence relations in RSM include both distance constraints and time constraints. The time constraints in RSM are easy to convert. Assume that there exists a minimum time constraint of SS with α lag time between activities M and N. Then, in each unit j, a forward arc with a positive value of α is used to connect the starting node of sub-activity M_j and the starting node of sub-activity N_j, as shown in Figure 4.3(a). Similarly, for other kinds of minimum time constraints (e.g., SF, FS, and FF), the conversion methods are shown in Figures 4.3(b)–(d), respectively.

For the distance constraints, it is necessary to match the sub-activities between pairs of activities first because the sub-activity of an activity in each unit activity may match with the sub-activity of the succeeding activity in a different unit. After that, the distance constraints in RSM could be represented by the precedence relations in the GPR network. As shown in Figure 4.4(a), there exists a minimum distance constraint between activities M and N with the constraint value of one unit. First, the sub-activity of activity M in each unit j is matched with that of activity N in unit $j-1$. Then a forward arc is used to connect sub-activities M_j and N_j, as shown in Figure 4.4(b).

The preceding steps yield a network model for which the earliest possible start schedule can be computed using the label-correcting

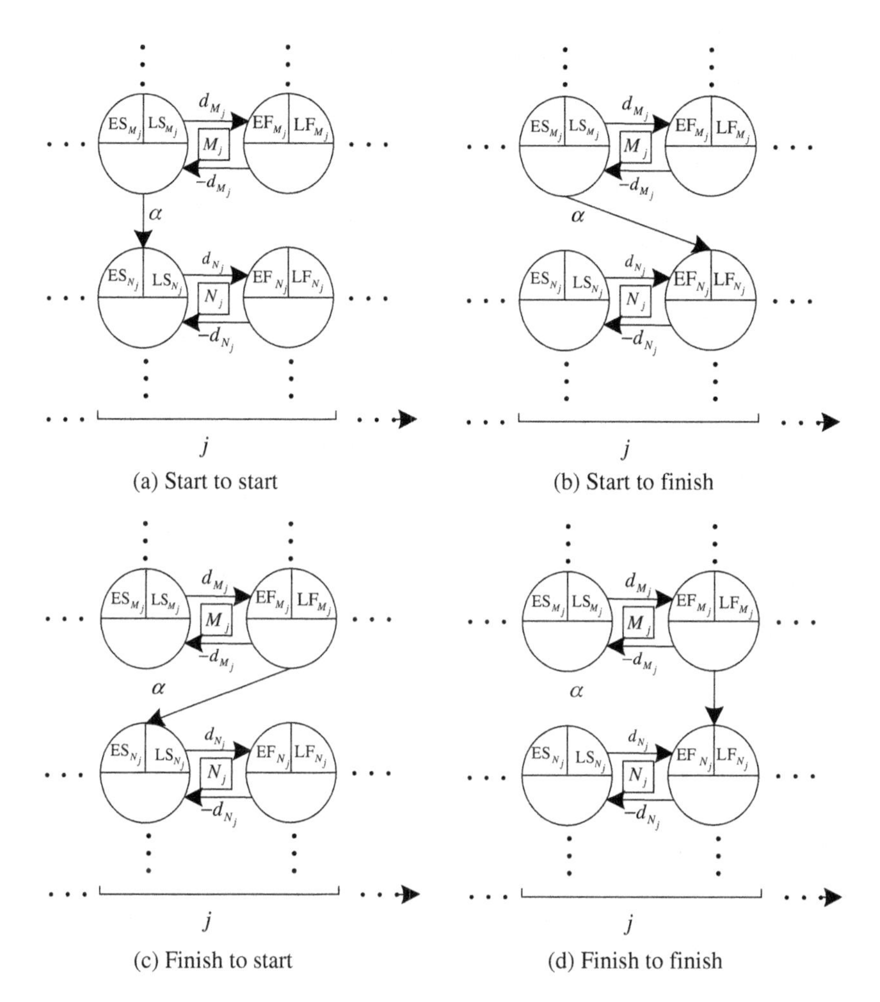

Figure 4.3 Conversion of time constraints.

algorithm (Ahuja et al., 1989), while the latest allowable start schedule can be similarly computed by reversing all arcs within the network under the condition that $es_n = ls_n$, where es_n and ls_n denote the earliest and latest start times of the finish node n, respectively. Through this process, the critical path is identified.

4.2.4 Displaying the Spatial Information

A horizontal axis is added at the bottom of the GPR network to represent the production unit. In an RSM schedule, the plane of the coordinate system helps to identify potential conflicts between two or more activities. Such a conflict may be a physical congestion where activities

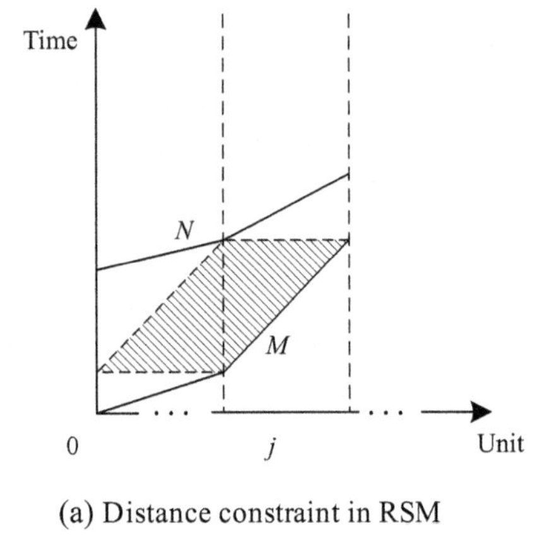

(a) Distance constraint in RSM

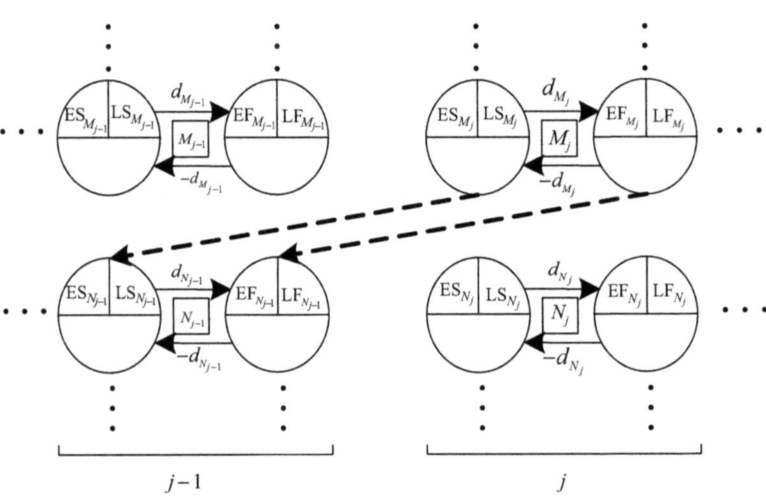

(b) Equivalent representation in GPRs

Figure 4.4 Conversion of distance constraints.

are in proximity to each other or an actual interference where activities are touching or crossing, as shown in Figure 4.5. In the GPR network, potential conflicts in space may occur in one of the following conditions: (1) $ES_{M_j} < ES_{N_j}$ and $EF_{M_j} > EF_{N_j}$; (2) $ES_{M_j} < LS_{N_j}$ and $EF_{M_j} > LF_{N_j}$; (3) $LS_{M_j} < ES_{N_j}$ and $LF_{M_j} > EF_{N_j}$; and (4) $LS_{M_j} < LS_{N_j}$. As all the start times and finish times of sub-activities have been calculated and shown in the GPR network, it can display the potential conflicts in space as RSM does, as well.

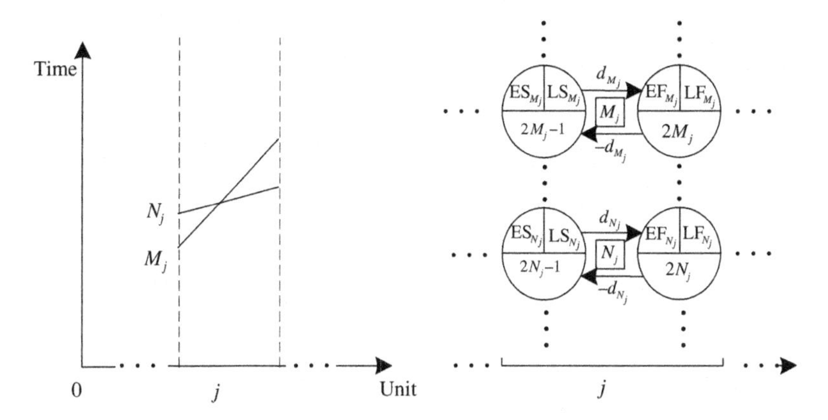

Figure 4.5 Display of potential conflicts in space.

Table 4.1 Information for the Illustrative Project

Activity/Unit	Duration (days)					Constraints	Value
	1	2	3	4	5		
Excavation (*A*)	3	3	3	5	5		
Lay pipe (*B*)	10	10	4	4	4	FF and SS	2 days
Test pipe (*C*)	1	1	1	1	1	Distance	2 units
Backfill (*D*)	9	8	8	8	8	FF and SS	3 days
Road reinstatement (*E*)	2	2	2	2	2	Distance	1 units

4.3 COMPARISON BETWEEN THE CONTROLLING PATH IN RSM AND THE CRITICAL PATH IN NETWORK MODEL

To compare the controlling path in RSM and critical path in the GPR network, a project involving a gas-pipe relocation is analyzed; this project was first presented by Kallantzis et al. (2007). The project consists of five units, and each includes the following activities in sequence: excavations, lay pipe, test pipe, backfill, and road reinstatement. Information on the project is shown in Table 4.1.

In order to highlight the effect of the resource continuity constraint on the controlling path, the following three cases are examined:

Case 1: All of the activities are required to maintain resource continuity.
Case 2: Except for activity *C*, there is no requirement for resource continuity; that is, only activity *C* cannot be interrupted.
Case 3: There is no requirement for resource continuity for any activity.

4.3.1 Analysis Based on Case 1

The RSM schedule for Case 1, where no activity is allowed to be interrupted, is shown in Figure 4.6. The controlling path is shown as the bold path, and it determines the project duration to be 77 days. For all activities, Table 4.2 lists the coordinates of preceding and succeeding controlling points (CPs), and the start times in the first unit and finish times in the last unit for Case 1.

Then, convert the RSM to the GPR network according to the proposed method. First, each sub-activity in RSM corresponds to an independent activity in the GPR network. Second, when dealing with the precedence relations, the sub-activities on activity *B* need to match their corresponding sub-activities on activity *C* because the constraint

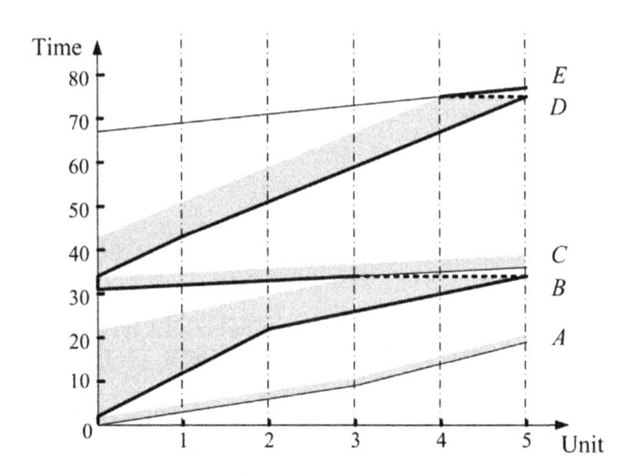

Figure 4.6 RSM diagram in Case 1.

Table 4.2 Calculation Results of the Controlling Path (Case 1)

Activity	Start Time in the First Unit (days)	Finish Time in the Last Unit (days)	Preceding CPs	Succeeding CPs
A	0	19	(0,0)	(0,0)
B	2	34	(0,2)	(5,34)
C	31	36	(3,34)	(0,31)
D	34	75	(0,34)	(5,75)
E	67	77	(4,75)	(5,77)

existing between activities B and C is the distance constraint of 2 units. Therefore, sub-activities B_3, B_4, and B_5 correspond to sub-activities C_1, C_2, and C_3, respectively. Similarly, the distance constraint existing between activity D and E makes sub-activities D_2, D_3, D_4, and D_5 correspond to sub-activities E_1, E_2, E_3, and E_4, respectively. And the precedence relation is SS and FF with minimum time constraint value 0. Third, when converting the logical relations, two arcs with opposite directions and time constraint equal to 0 are used between logical adjacent activities in the GPR network because all the activities are required to maintain the resource continuity in RSM. The critical path is shown as the bold path in Figure 4.7. It also determines the project duration as 77 days. The network model coincides with RSM completely.

4.3.2 Analysis Based on Case 2

In Case 2, there is no requirement for resource continuity for any activity except C. The RSM diagram for this is shown in Figure 4.8, with a corresponding GPR network shown in Figure 4.9. Table 4.3 lists the start times in the first unit, finish times in the last unit, and coordinates of CPs for all the five activities for Case 2. Note that while converting the logical relations, only forward arcs with value 0 are used in activities A, B, D, and E, but both forward and backward arcs with time constraints equal to 0 are used in activity C. The durations of both of the schedules are 77 days, and the GPR network coincides with RSM completely.

4.3.3 Analysis Based on Case 3

In Case 3, there is no resource continuity constraint for all activities; that is, all activities are allowed to be interrupted. The RSM schedule and the corresponding GPRs network are shown in Figures 4.10 and 4.11, respectively. Table 4.4 lists the start times in the first unit, finish times in the last unit, and coordinates of CPs for all five activities for Case 3. Note that in converting the logical relations for all activities, only forward arcs with time constraints equal to 0 are used, as there is no requirement for resource continuity. The duration of both of the schedules is 71 days, with the GPR network once again coinciding completely with RSM.

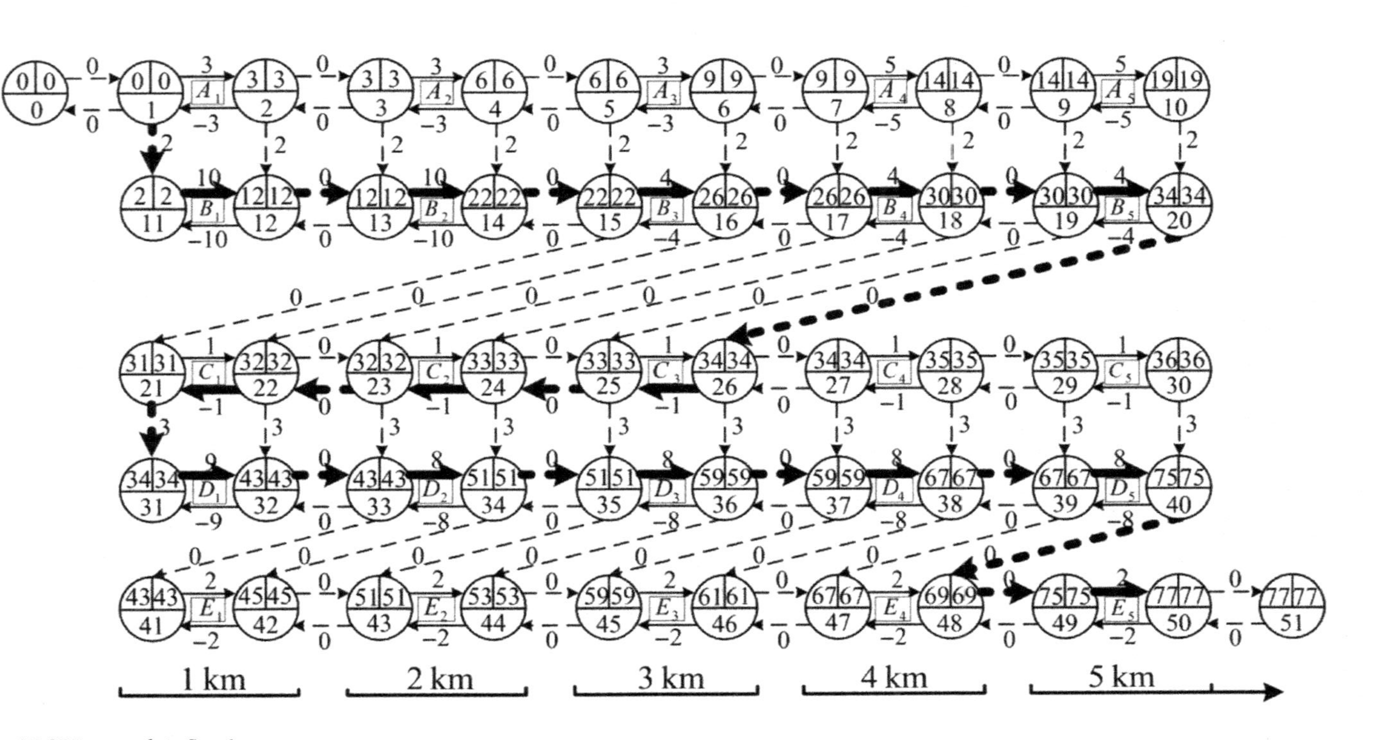

Figure 4.7 GPRs network in Case 1.

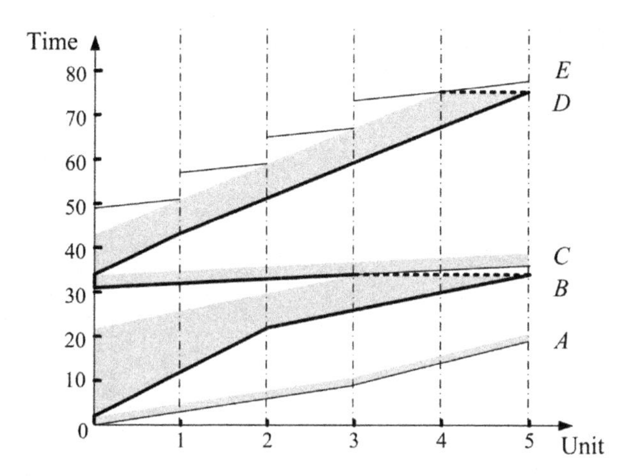

Figure 4.8 RSM diagram in Case 2.

4.3.4 Discussion

If we compare these results with the study by Kallantzis et al. (2007), two major differences are apparent. First, the representations of resource continuity are different. In this chapter, a backward arc with zero value is used to maintain resource continuity. In the study by Kallantzis et al. (2007), there is no way to keep resource continuity for the activities. According to their method, the projects whose activities are different in resource continuity requirement will correspond to the same network model. For instance, the RSM in Case 1 (Figure 4.6) was converted to the network in Case 3 (Figure 4.11).

Second, relationships between RSM and the network model are different. In this chapter, the RSM coincides with the network model in makespan, critical path, and activity criticality in each case. In the study by Kallantzis et al. (2007), only the project in Case 1 was analyzed, and RSM was different from the network model.

4.4 COMPARISON OF ACTIVITY CRITICALITY

Elmaghraby and Kamburowski (1992) partitioned all critical activities of networks under generalized precedence relations into five categories. For RSM, Zhang and Qi (2012) defined three types of controlling segments, namely the forward controlling, point controlling, and backward controlling segments.

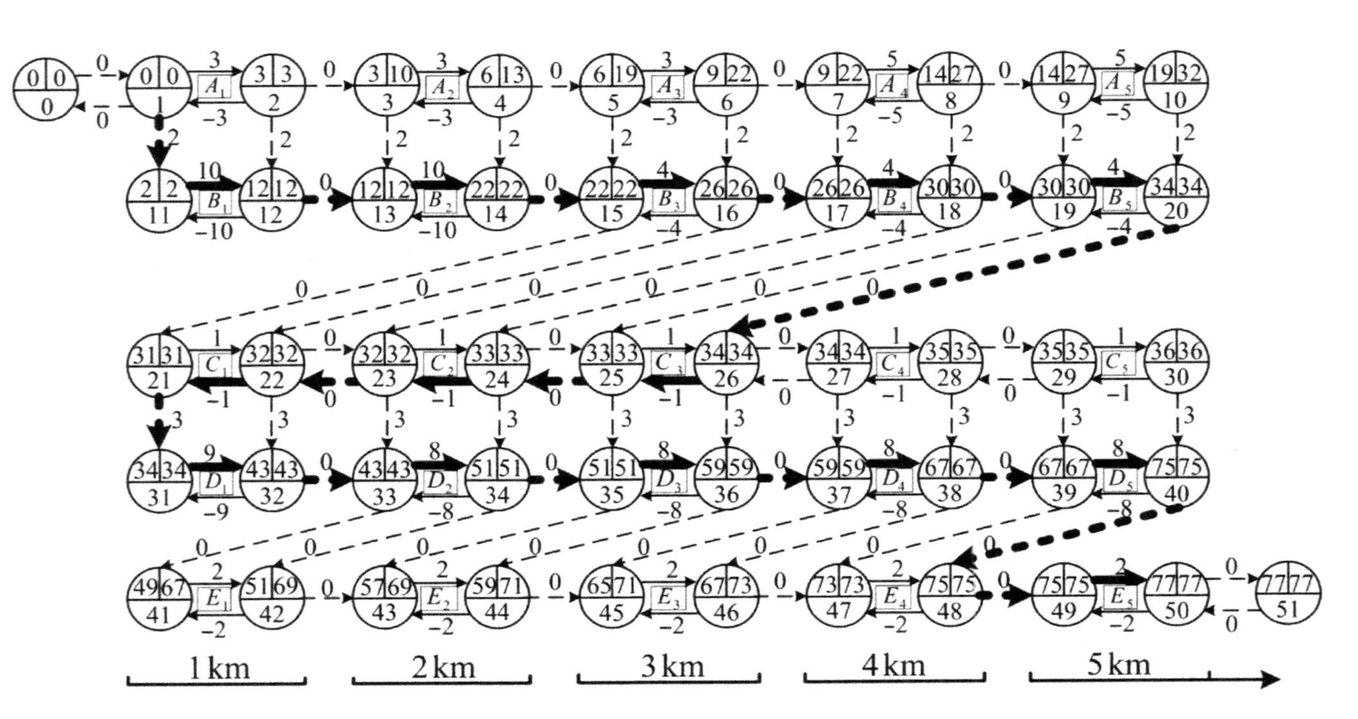

Figure 4.9 GPRs network in Case 2.

Table 4.3 Calculation Results of the Controlling Path (Case 2)

Activity	Start Time in the First Unit (days)	Finish Time in the Last Unit (days)	Preceding CPs	Succeeding CPs
A	0	19	(0,0)	(0,0)
B	2	34	(0,2)	(5,34)
C	31	36	(3,34)	(0,31)
D	34	75	(0,34)	(5,75)
E	49	77	(4,75)	(5,77)

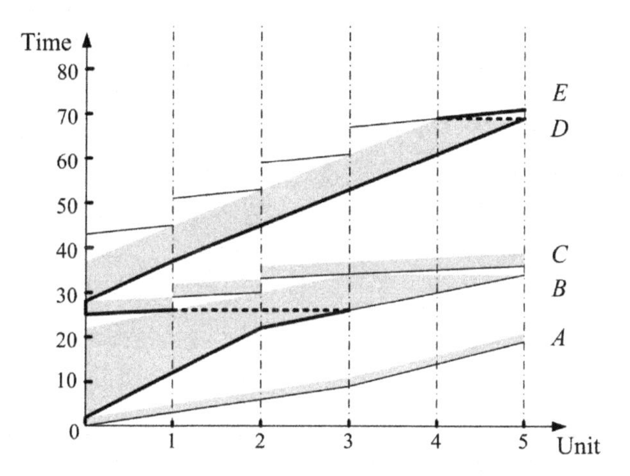

Figure 4.10 RSM diagram in Case 3.

The forward controlling segment is similar to the forward critical activity of a network, and the project duration will change in the same direction as that of the forward controlling segment; for instance, activities *B* and *D* and sub-activity E_5 are forward controlling segments in Figure 4.6 that correspond to the forward critical activities in Figure 4.7.

The point controlling segment corresponds to the start-critical or finish-critical activities of the network. For instance, the starting point of sub-activity A_1 in Figure 4.6 is a point controlling segment, while in Figure 4.7, activity A_1 is a start-critical activity.

The backward controlling segment in RSM corresponds to the backward critical activity in a network model; for example, sub-activities C_1, C_2, and C_3 in Figures 4.6 and 4.8 and sub-activity C_1 in Figure 4.10 are all backward controlling segments. The backward

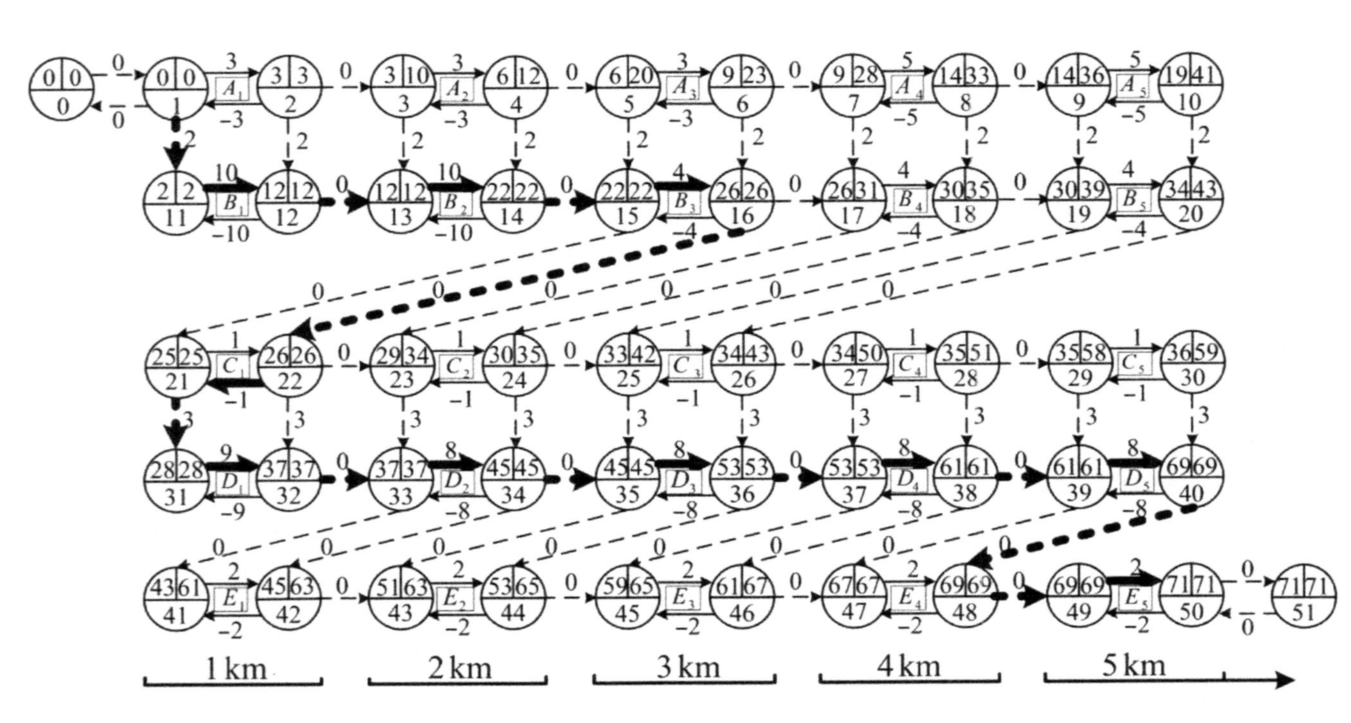

Figure 4.11 GPRs network in Case 3.

Table 4.4 Calculation Results of the Controlling Path (Case 3)

Activity	Start Time in the First Unit (days)	Finish Time in the Last Unit (days)	Preceding CP	Succeeding CP
A	0	19	(0,0)	(0,0)
B	2	34	(0,2)	(3,26)
C	25	36	(1,26)	(0,25)
D	28	69	(0,28)	(5,69)
E	43	71	(4,69)	(5,71)

controlling segment is the key to explaining the reason that the controlling path in RSM does not coincide with the critical path in the network by Kallantzis et al. (2007). When the backward controlling segment containing two or more sub-activities in RSM is converted to the network model without maintaining resource continuity, it will be split into several activities with interruptions between them during the process of computing the earliest possible start schedule and the latest allowable start schedule. For example, according to the method proposed by Kallantzis et al. (2007), the backward controlling segments C_1 through C_3 in Cases 1 (Figure 4.6) and 2 (Figure 4.8) are converted to activities C_1, C_2, and C_3 in Case 3 (Figure 4.10); however, C_2 and C_3 are not included in the critical path in computing the minimum longest path (the critical path) when resource continuity is not maintained. In other words, in Cases 1 (Figure 4.6) and 2 (Figure 4.10), the start time of activity C_1 is determined by activities C_2 and C_3, in turn, because of the resource continuity constraint on activity C, while C_3 is determined by activity B_5. In Case 3, where there is no resource continuity constraint, C_1 is determined by B_3, which is 6 days earlier, and therefore the project duration is shortened by 6 days when RSM is converted to the network.

If there is no backward controlling segment containing two or more sub-activities, this discrepancy will not occur. For example, in project 11 (one of the 25 random projects presented by Kallantzis et al. (2007) and in the two projects presented by Yamin and Harmelink (2001), the controlling paths in RSM coincide with the critical paths in the network model owing to the lack of backward controlling segments.

In general, these results show that the criticality of RSM coincides with that of its corresponding GPR network model.

4.5 CONCLUSION AND PROSPECTS

This chapter has presented a method for converting RSM to a network model (i.e., GPR network). In the existing literature, the conversion of an RSM to a network model may alter the makespan and the critical path, which may confuse researchers. The method proposed in this chapter, on the other hand, guarantees conservation of both makespan and criticality. A comparison of the criticality between the RSM and the network model has shown that the production of altered makespans and criticalities in previous methods results from the conversion of backward controlling segments containing two or more subactivities, without maintaining resource continuity. This finding helps researchers to clear up confusion in understanding the relationship between the controlling path in RSM and the critical path in the network model. Using the method proposed in this chapter, RSM will completely coincide with its corresponding network in terms of criticality.

Both the RSM and the network model have advantages in scheduling repetitive projects. Practitioners can use RSM to manage both time and space on the project site in a graphical display. The network model, on the other hand, is more commonly accepted by owners and construction contractors. This chapter offers practitioners a method for converting an RSM to a network model easily and accurately, and by using this method, the RSM work continuity requirement can be maintained and the distance constraint accurately converted into the network model. Practitioners can realize the benefits of both methodologies by using RSM as a tool for planning and controlling and then converting RSM to a network model when required contractually. In this way, the RSM will exploit its advantages in scheduling repetitive projects and be accepted by more practitioners.

This chapter focuses on the conversion method and comparison of criticality. It will help us to engage in some other important issues in RSM including resource management, floats analysis, and applications, especially the correspondence between the RSM and network model. Those will be our future studies.

CHAPTER 5

Resource Allocation Problem in Repetitive Construction Projects

5.1 INTRODUCTION

Solving the resource allocation problem in repetitive construction projects aims at minimizing project duration by determining the optimum execution modes and start times (or interruption strategies) for all sub-activities, while satisfying a constraint system that may consist of precedence relations, logical relations, and resource continuity. This type of problem is more complicated and harder to solve than those in nonrepetitive projects, since the resource continuity constraint and multiple types of time constraints must be considered in the optimization process.

Available planning and scheduling models that focus on minimizing the duration of repetitive construction projects can be grouped into two main categories: (1) models that provide strict compliance with the resource continuity constraint; and (2) models that allow interruptions in crew work continuity.

Selinger (1980) presented a dynamic programming algorithm for this problem, which takes execution modes of activities as decision variables and emphasizes that all activities must satisfy the resource continuity constraint. Despite the apparent advantages of maintaining resource continuity (maximization of the learning curve effect and minimization of idle time of each crew), its strict application may lead to longer overall project duration. Therefore, the author further suggested that the violation of the resource continuity constraint, allowing work interruptions, might reduce overall project duration. Russell and Caselton (1988) extended the work of Selinger in developing a two-state variable, *N*-stage dynamic programming model. The two state variables are vectors, with the first representing a set of possible durations of activities, and the second representing a set of interruption durations between different units of the same activity. However, this model requires the planners to arbitrarily specify, prior to scheduling,

Repetitive Project Scheduling: Theory and Methods.

a set of interruption vectors for each activity. Such a requirement is not practical and may render the optimization problem infeasible. To circumvent the limitations, El-Rayes and Moselhi (2001) presented an automated optimization model. This model utilizes a dynamic programming formulation and incorporates a scheduling algorithm and an interruption algorithm so as to automate the generation of interruptions during scheduling. In many cases, these methods are not easy to deal with in complex projects due to enormous numbers of decision variables. Hence, heuristic methods have been developed for scheduling repetitive construction projects in practice. Hyari and El-Rayes (2006) presented a genetic algorithm (GA)–based multi-objective optimization model for minimizing project duration and maximizing continuity of resources. This model is organized in three major modules: scheduling, optimization, and ranking modules. First, the scheduling module uses a resource-driven scheduling algorithm to develop practical schedules for repetitive construction projects. Second, the optimization module utilizes multi-objective GAs to search for and identify feasible construction plans that establish optimal trade-offs between project duration and interruption days. Third, the ranking module uses multi-attribute utility theory to rank the generated plans in order to facilitate the selection and execution of the best overall plan for the project being considered. Liu and Wang (2007) developed a flexibility model involving different objectives and resource assignment tasks. Their model adopted constraint programming (CP) as the searching algorithm for model formulation; the CP program creates the flexibility for optimizing either total cost or project duration. Additionally, the concept of outsourcing resources is introduced to improve project performance. It should be noted that the optimization strategy of all the above heuristic methods is to search for the optimal solution in all possible combinations of execution modes of activities and interruption durations between different units of the same activity. This may result in premature convergence for large-scale problems.

Long and Ohsato (2009) developed a GA method for scheduling repetitive construction projects with several objectives such as project duration, project cost, or both. This method considers different attributes of activities (such as activities which allow or do not allow interruption) to provide a satisfactory schedule. In order to minimize the objectives, the proposed method finds a set of suitable durations for activities by GA, and then determines the suitable start times for these activities by a scheduling algorithm. This means the interruption durations between units of the same activity are no longer the decision variables.

In RSM, the length of the controlling path determines project duration. There is a positive correlation between the duration of forward controlling segments and project duration, and a negative correlation between the duration of backward controlling segments and project duration. This chapter presents a fast optimization algorithm to solve the resource allocation problem in repetitive construction projects.

5.2 PROBLEM FORMULATION

Consider a repetitive construction project consisting of I activities and J units; a_{ij} ($i = 1,2,\ldots,I$; $j = 1,2,\ldots,J$) denotes the sub-activity of activity i in unit j. Sub-activities a_{11} and a_{IJ} represent the start and finish of the project, respectively. The hard logic assumption is adopted and the fixed sequence from unit 1 to unit J is assumed for all activities. Activity i has K_i alternative execution modes, and all sub-activities in the same activity must be performed by the same mode. d_{ijk} denotes the duration of sub-activity a_{ij} in mode k. Each pair of activities, (i,l) in sets E(SS), E(SF), E(FS), and E(FF), should satisfy the minimum time constraint of SS, SF, FS, and FF, respectively, where activity l is the succeeding activity of i. The variable s_{ij} denotes the start time of sub-activity a_{ij}. Binary variables x_{ik} take value one, if activity i is performed by mode k, and zero otherwise. Finally, the resource allocation problem can be described by the following integer linear programming model:

$$\text{Min } s_{IJ} + \sum_{k=1}^{K_I} x_{Ik} d_{IJk} \tag{5.1}$$

such that

$$s_{ij} + \sum_{k=1}^{K_i} x_{ik} d_{ijk} \le s_{i(j+1)}, \quad i \in w_1, \quad j = 1,\ldots,J-1 \tag{5.2}$$

$$s_{ij} + \sum_{k=1}^{K_i} x_{ik} d_{ijk} = s_{i(j+1)}, \quad i \in w_2, \quad j = 1,\ldots,J-1 \tag{5.3}$$

$$s_{ij} + t_{il} \le s_{lj}, \quad (i,l) \in E(\text{SS}), \quad j = 1,\ldots,J \tag{5.4}$$

$$s_{ij} + t_{il} \le s_{lj} + \sum_{k=1}^{K_l} x_{lk} d_{ljk}, \quad (i,l) \in E(\mathrm{SF}), \quad j = 1, \ldots, J \tag{5.5}$$

$$s_{ij} + \sum_{k=1}^{K_i} x_{ik} d_{ijk} + t_{il} \le s_{lj}, \quad (i,l) \in E(\mathrm{FS}), \quad j = 1, \ldots, J \tag{5.6}$$

$$s_{ij} + \sum_{k=1}^{K_i} x_{ik} d_{ijk} + t_{il} \le s_{lj} + \sum_{k=1}^{K_l} x_{lk} d_{ljk}, \quad (i,l) \in E(\mathrm{FF}), \quad j = 1, \ldots, J \tag{5.7}$$

$$\sum_{k=1}^{K_i} x_{ik} = 1, \quad x_{ik} = \{0, 1\}, \quad i = 1, \ldots, I \tag{5.8}$$

Constraints (5.2) and (5.3) guarantee that all activities are executed by the fixed sequence from unit 1 to unit J, and do not violate the given resource continuity constraint: each activity in set w_1 is allowed to be interrupted, and each activity in set w_2 must be performed without interruption. Constraints (5.4)–(5.7) guarantee the precedence relation constraints among activities, where t_{il} denotes the lag time between activity i and its succeeding activity l. Constraint (5.8) ensures that exactly one execution mode is chosen for each activity.

5.3 OPTIMIZATION METHOD BASED ON BACKWARD CONTROLLING SEGMENTS

In RSM, let A^+, A^- and $\overline{C}$ denote the forward controlling segment, backward controlling segment, and controlling constraint on the controlling path, respectively. According to Eq. (3.5), the project duration of a repetitive construction project can be estimated by

$$T = \sum_{a_{ij} \in A^+} d_{ij} - \sum_{a_{ij} \in A^-} d_{ij} + \sum_{(i,l) \in \overline{C}} t_{il} = \sum_{a_{ij} \in A^+} \sum_{k=1}^{K_i} x_{ik} d_{ijk} - \sum_{a_{ij} \in A^-} \sum_{k=1}^{K_i} x_{ik} d_{ijk} + \sum_{(i,l) \in \overline{C}} t_{il} \tag{5.9}$$

To obtain the shortest project duration, we first assume all activities select the fastest execution modes and all sub-activities start at their earliest times. As a result, an initial feasible schedule is generated, and its duration cannot be further reduced by increasing the productivities of activities. Then we identify the controlling path of this schedule. If there is no backward controlling segment on the controlling path, according to Eq. (5.9), the schedule obtained is the optimal solution for the resource allocation problem. Otherwise, the only way to further shorten project duration is to prolong the durations of some backward controlling segments. This may result in the generation of new backward controlling segments. If we can find the prerequisites that a segment can be a backward controlling segment, they can be used to simplify the current problem by performing all activities without meeting the prerequisites by their fastest execution modes and starting all sub-activities at their earliest start times. At present, the decision variables only are the execution modes of those activities that satisfy the prerequisites. Now the question is, what are the prerequisites?

Theorem 5.1: If there exists a backward controlling segment on activity *i*, this activity must satisfy one of the following two conditions:

1. There exists a minimum time constraint of SF or FF between activity *i* and its preceding activity, and a minimum time constraint of SF or in between activity *i* and its succeeding activity.
2. Activity *i* is required to meet the resource continuity constraint.

Proof: 1. If activity *i* is allowed to be interrupted.
Since there exists a backward controlling segment on activity *i* and this activity is allowed to be interrupted, the controlling path only passes one sub-activity of activity *i*, and this sub-activity is a backward controlling sub-activity. (This conclusion can easily be deduced by comparing the RSM schedule and its equivalent GPR network according to the analysis in Chapter 4.) Without loss of generality, we can assume that this sub-activity is a_{ij}. According to the definition of backward controlling segment, the preceding controlling point should be realized later than the succeeding controlling point. Then the preceding controlling point of activity *i* should be the ending point of a_{ij}, and its succeeding controlling point should be the starting point of a_{ij}. To satisfy this condition, activity *i* and its preceding activity must have a minimum time constraint of SF or FF; meanwhile, activity *i* and its succeeding activity must have a minimum time constraint of SF or SS.

2. If there is no minimum time constraint of SF or FF between activity i and its preceding activity, or no minimum time constraint of SF or SS between activity i and its succeeding activity.
At this point, the preceding controlling point of activity i must be the starting point of one sub-activity a_{ij}, and the succeeding controlling point of this activity must be the ending point of one sub-activity $a_{ij'}$. And the start time of a_{ij} should be greater than the finish time of $a_{ij'}$. This means that these two sub-activities cannot be the same sub-activity, and $j > j'$. To guarantee that the controlling path passes unit j earlier than unit j', the resource continuity constraint should be maintained strictly by activity i.

5.4 PROPOSED GENETIC ALGORITHM

Based on the analysis in the above section, the improved method for the resource allocation problem in repetitive construction projects is proposed below. This method adopts GA as the searching algorithm, in which PARENT and CHILD denote the parent and offspring populations, respectively.

Step 1. For the project to be optimized, classify all activities that satisfy the prerequisites (see in Theorem 5.1) into set ξ; the rest are classified into set ζ. All activities in set ζ are performed by their fastest execution modes and all sub-activities are scheduled at their earliest start times.
Step 2. If $\xi = \varnothing$, terminate the procedure and return to the optimal solution; otherwise go to the next step.
Step 3. Encode the execution modes of activities in set ξ to generate the initial population PARENT consisting of N_p individuals. Calculate the objective function value of each individual by decoding its chromosome and starting all sub-activities at their earliest times.
Step 4. Evaluate the fitness of each individual in PARENT using the reciprocal of the objective function value. Then create the offspring population CHILD by using roulette selection, single-point crossover, and single-point mutation operations.
Step 5. Calculate the fitness of each individual in CHILD, and recombine PARENT = selected individuals from PARENT and CHILD. If the terminating condition is met, return to the best found solution. Otherwise, go back to Step 4.

5.5 CASE STUDY

5.5.1 An Illustrative Example

A fictitious project is analyzed first. This project consists of five units with the input data shown in Table 5.1. Each unit includes 14 activities as shown in Figure 5.1. This is a typical project, and therefore the quantities of work of each activity in different units are always the same. The project manager requires that activities 2, 4, and 11 must be performed without interruption for the purpose of saving on cost. Now we need to calculate the shortest duration of this project.

Because activities 2, 4, 7, and 11 satisfy the prerequisites that a segment can be a backward controlling segment, only the execution modes of these activities need to be taken as the decision variables. For comparison purposes, we use the GAs presented by Long and Ohsato (2009), Hyari and El-Rayes (2006), and in this chapter to solve this problem. As shown in Figure 5.2, although the shortest project duration calculated by all the methods is 109 days, the proposed method takes less CPU time than the other two methods to find the optimal solution.

Table 5.1 Project Information

Activity	Unit Duration (days)		
	Mode 1	Mode 2	Mode 3
1	0	0	0
2	24	20	16
3	20	18	—
4	5	4	3
5	10	9	8
6	5	4	3
7	12	10	8
8	10	7	—
9	2	—	—
10	2	—	—
11	4	3	2
12	3	2	—
13	4	3	—
14	0	0	0

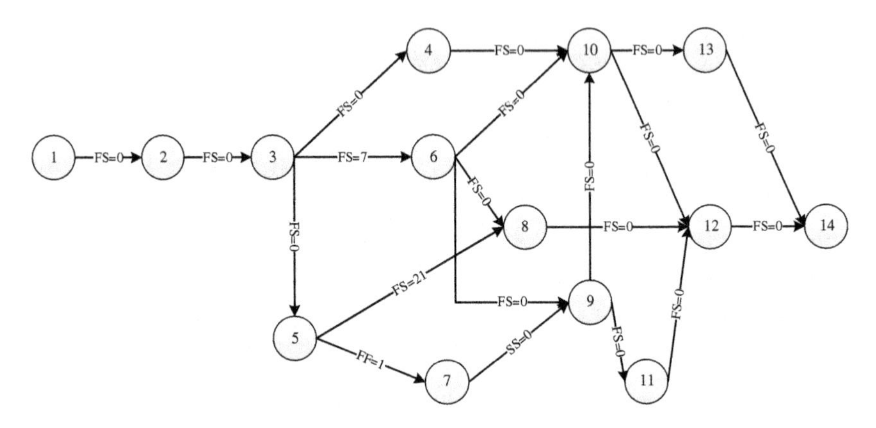

Figure 5.1 AON of the example project.

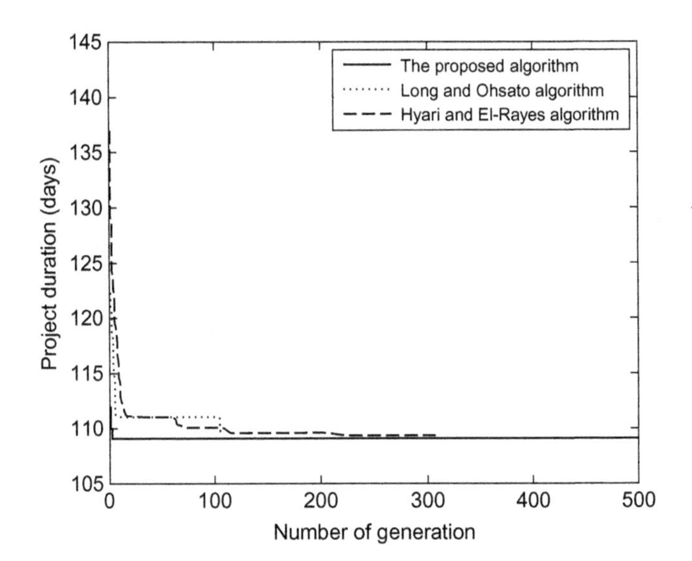

Figure 5.2 Rates of convergence of different algorithms.

To further compare the performance of different algorithms, this chapter makes a more comprehensive analysis through the following random test problems.

5.5.2 Test Problems

The experiments are performed on a personal computer with a 2.00 GHz CPU. All the methods to be tested are compiled with MATLAB 7.11 and tested under Windows 7 Professional. Since there are no available open test instances for the resource allocation problem in repetitive construction projects, we randomly generate the test instances by the following steps:

Step 1. Determine randomly the problem scale, denoted by $A(I)U(J)$, where I and J denote the number of activities and units, respectively.
Step 2. Generate randomly the set of successors for each activity; determine the type and amount of minimum time constraint between activity i and each of its succeeding activities. Then randomly determine whether the resource continuity constraint should be satisfied for each activity i.
Step 3. Randomly select a positive integer K_i within [1, 6] to represent the number of alternative execution modes of activity i. Then the unit duration of activity i in mode k, d_{ijk}, is assigned a random integer within [1, 10]. Note that the unit durations of an activity in any two modes cannot take the same value.

5.5.3 Population Size, Crossover and Mutation Probability

In GA, on the one hand, a high crossover and mutation probability imply that these two operators are frequently used, and this may induce a high time consumption over generations; on the other hand, it is evident that a large population size requires more computational effort than a low one. In addition, the number of generations performed depends on the computational effort and on the given time limit. In fact, great population size, crossover, and mutation probability require a long computation; so, for limited CPU time, the algorithm computes a small number of schedules. Consequently, we choose to experiment with population size (N_p), crossover probability (P_c), and mutation probability (P_m) on the basis of a 1 s stopping criterion. The average deviations from the optima are reported in Table 5.2. The investigation of population size, crossover, and mutation probability leads to best results with $P_c = 0.7$ and $P_m = 0.5$ when $N_p = 30$, and $P_c = 0.9$ and $P_m = 0.5$ when $N_p = 60$.

5.5.4 Algorithms Comparison

Taking the fastest CPU time as the test metric, the calculation results for each algorithm under different scales of problems are listed in Table 5.3.

Table 5.2 Average Deviations: $A(20)U(10)$, 1 s

$N_p = 30$	$P_c = 0.5$	$P_c = 0.7$	$P_c = 0.9$
	(%)		
$P_m = 0.1$	2.06	2.06	2.21
$P_m = 0.3$	1.55	1.90	1.72
$P_m = 0.5$	1.69	**1.46**	1.58
$N_p = 60$	(%)		
$P_m = 0.1$	0.9	1.06	1.16
$P_m = 0.3$	0.9	0.98	1.01
$P_m = 0.5$	1.02	1.00	**0.8**

Note: The optimal solution is obtained by using LINGO.

Table 5.3 Fastest CPU Time

Test Problems	B&B[1]	B&B[2]	Proposed GA (60,0.9,0.5)	Proposed GA (30,0.7,0.5)	Long and Ohsato GA (60,0.9,0.5)	Long and Ohsato GA (30,0.7,0.5)
$A(10)U(10)$	1.3000	1.0300	0.0001	0.0001	0.0437	0.1279
$A(15)U(10)$	8.8500	5.4000	0.0124	0.0032	0.4585	1.8778
$A(20)U(10)$	65.6000	25.4500	0.0125	0.0118	1.7556	2.3217
$A(25)U(10)$	113.7000	71.2500	0.0321	0.0492	3.0456	4.4219
$A(25)U(15)$	163.2000	89.0000	0.0484	0.0718	3.0888	8.8903

Note: B&B[1] denotes a linear model in which the execution modes of all activities and the start times of all sub-activities are taken as the decision variables. B&B[2] denotes another linear model where only the execution modes of those activities that satisfy the prerequisites listed in Theorem 5.1 are taken as the decision variables. Both B&B[1] and B&B[2] are run on a LINGO compiler.

As the problem scale expands, the differences in fastest CPU time between B&B[1] and B&B[2] increases. The same trend occurs in the comparison between the proposed GA and the GA presented by Long and Ohsato (2009). For different scales of test problems, the average deviations from the optima under a 1 s stopping criterion are listed in Table 5.4, where the average deviation calculated by the proposed algorithm is always less than that calculated by the other two methods.

When the CPU time is limited to 10 s, the likelihood of finding the optimal solution for different algorithms is shown in Table 5.5. From the results obtained, the GA presented by Hyari and El-Rayes (2006) is unable to obtain the optimal solution for any one test problem within 10 s; the GA presented by Long and Ohsato (2009) is only guaranteed to obtain the optimal solution with a high probability for

Table 5.4 Average Deviations. 1 s

Test Problems	Proposed GA (60,0.9,0.5)	Proposed GA (30,0.7,0.5)	Long and Ohsato GA (60,0.9,0.5)	Long and Ohsato GA (30,0.7,0.5)	Hyari and El-Rayes GA (60,0.9,0.5)	Hyari and El-Rayes GA (30,0.7,0.5)
	(%)					
$A(10)U(10)$	0	0.20	2.43	5.48	17.69	22.30
$A(15)U(10)$	0.03	0.37	6.74	14.57	33.13	46.21
$A(20)U(10)$	0.8	1.46	11.45	18.09	43.04	58.84
$A(25)U(10)$	0.27	0.57	14.51	19.58	52.56	68.96
$A(25)U(15)$	1.35	2.64	25.57	35.14	72.76	92.94

Table 5.5 Likelihood of Finding the Optimal Solution: 10 s

Test Problems	Proposed GA (60,0.9,0.5)	Proposed GA (30,0.7,0.5)	Long and Ohsato GA (60,0.9,0.5)	Long and Ohsato GA (30,0.7,0.5)	Hyari and El-Rayes GA (60,0.9,0.5)	Hyari and El-Rayes GA (30,0.7,0.5)
	(%)					
$A(10)U(10)$	100	100	89.00	78.00	0	0
$A(15)U(10)$	98.50	99.17	50.00	42.00	0	0
$A(20)U(10)$	92.00	91.67	36.83	17.50	0	0
$A(25)U(10)$	84.33	86.50	7.33	4.00	0	0
$A(25)U(15)$	80.67	85.33	2.17	1.67	0	0

small-scale problems. The proposed algorithm shows a considerable advantage over the other two methods, because it is very likely to get the optimal solution within 10 s even for the large-scale problems. The results of the comparisons presented above, despite the limited scope of the analysis, support the authors' thesis: using the characteristic of backward controlling segments can simplify the resource allocation problem in repetitive construction projects.

5.6 CONCLUSION AND PROSPECTS

Based on the fact that the duration of forward controlling segments is proportional to project duration, and the duration of backward controlling segments is inversely proportional to project duration, this chapter has presented a fast optimization algorithm to solve the resource allocation problem in repetitive construction projects. This algorithm only considers the execution modes of those activities which satisfy the prerequisites for a segment to be a backward controlling segment as the decision variables, while other activities are performed by their fastest execution modes. Compared with existing algorithms, the proposed method requires less CPU time.

Backward controlling segments are very common in repetitive construction projects. A suitable future endeavor is to explore the application of backward controlling segments to the time/cost trade-off problem in repetitive construction projects.

CHAPTER 6

Resource-Constrained Scheduling in Repetitive Construction Projects

6.1 INTRODUCTION

The resource-constrained project scheduling problem (RCPSP) has been a research topic for many decades, resulting in a wide variety of optimization procedures that differ in objective functions, activity assumptions, resource constraints, and many other aspects. In general, the objective of RCPSP is to minimize the total duration or makespan of a project subject to precedence relations between activities and limited renewable resource availability. The RCPSP is known to be an Non-deterministic Polynomial-time hard (NP-hard) problem (Blazewicz et al., 1983). Extensions to other objective functions (e.g., net present value maximization), resource constraints (to nonrenewable and doubly constrained resources), and multiple activity modes often result in highly complex optimization problems.

Slowinski (1980) categorized resources used by project activities as renewable, nonrenewable, and doubly constrained. Renewable resources are periodically renewed, but their quantity is limited over each time period and may differ from one period to the next. Some examples are manpower, machines, equipment, power, and fuel flow. For nonrenewable resources, constraints on availability only concern total consumption over the whole period of project duration and not at each time period. Raw materials are a typical example of nonrenewable resources, since they are available in a specific quantity for a project. Doubly constrained resource quantities are constrained both per period and per project. Money is an example of such a resource, since there is usually a specific total budget for the entire project, as well as a limited cash flow per period, according to progress. As formally shown by Talbot (1982), each doubly constrained resource can be represented by one renewable and one nonrenewable resource, respectively.

Repetitive Project Scheduling: Theory and Methods.

Many research efforts have extended the RCPSP to the presence of multiple activity modes, where each activity can be executed over a different duration and a corresponding renewable and nonrenewable resource use. In this chapter, we consider the multi-mode resource-constrained project scheduling problem (MRCPSP) with only renewable resource constraints in repetitive construction projects. To have a comprehensive understanding of RCPSP and MRCPSP, please refer to the overview papers of Icmeli et al. (1993), Brucker et al. (1999), Herroelen et al. (1998), Hartmann and Kolisch (2000), and Kilisch and Hartmann (2006), and the research handbook by Demeulemeester and Herroelen (2002). Here we examine only solution methods for resource-constrained scheduling in repetitive construction projects.

Leu and Hwang (2001) developed a genetic algorithm–based method for precast production with consideration of resource constraints and resource sharing. The line-of-balance technique is applied to scheduling precast production, in which work interruption and single mode are considered. Hsie (2009) presented an optimization model based on an evolutionary strategy algorithm for MRCPSP in linear projects, in which all activities must keep strict continuity of resources. The fundamental assumption is for crews to maintain the same production rate in each time period rather than each production unit in space segments; that is, there is no need for changing tools, extra preparation, or warming up in the middle of a time period. The reason the author adopted this assumption is that a crew maintaining the same production rate within each space segment may render the schedule inefficient when the length of the space segment is invisible to the production rate. In this case, a crew would need to change its size, composition, or associated equipment in the middle of a time period, causing unproductive preparation and warm-up. Another reason is that production units in space segments cannot be directly linked by the daily or weekly payment schedule, thus involving extra administrative work. Unlike the existing studies, this chapter takes both multiple modes and work interruption into consideration; the single crew assumption is also adopted.

6.2 PROBLEM DESCRIPTION

Consider a repetitive construction project that includes I activities labeled $i = 1,2, \ldots, I$ and J units labeled $j = 1,2, \ldots, J$. Activity 1 is the only start activity and activity I is the only finish activity. Each sub-activity may be executed in one of several modes. A mode is a way of performing a job. It reflects, for the sub-activity in question, first, the consumption of each resource, and second, the related duration. Each sub-activity a_{ij}, $i = 1,2, \ldots, I$, $j = 1,2, \ldots, J$, has K_i execution modes. The sub-activity a_{ij} performed on mode k, $k = 1, \ldots, K_i$, has a processing time referred to as d_{ijk}; it requires r_{ikl} units of each renewable resource l, $l = 1,2, \ldots, L$ in each period it is processed. The maximal availability of each renewable resource l in each period is R_l. The precedence relation between activities in each unit of finish to start with zero lag time is considered. A schedule is defined by vectors of activity start times and modes in all units; it is said to be feasible if all precedence relations, logical relations, and renewable resource constraints are satisfied. The objective of the problem type is to find a feasible schedule with the lowest possible project duration and that can be formulated mathematically as follows:

$$\min \sum_{t=\mathrm{ES}_{IJ}}^{\mathrm{LS}_{IJ}} \sum_{k=1}^{K_I} (t + d_{IJk}) x_{IJkt} \tag{6.1}$$

Subject to

$$\sum_{k=1}^{K_i} \sum_{t=\mathrm{ES}_{ij}}^{\mathrm{LS}_{ij}} x_{ijkt} = 1, \quad i = 1, \ldots, I;\ j = 1, \ldots, J \tag{6.2}$$

$$\sum_{k=1}^{K_i} \sum_{t=\mathrm{ES}_{ij}}^{\mathrm{LS}_{ij}} (t + d_{ijk}) x_{ijmt} \le \sum_{k=1}^{K_s} \sum_{t=\mathrm{ES}_{sj}}^{\mathrm{LS}_{sj}} t x_{sjmt}, \quad (i, s) \in E_{\mathrm{FS}};\ j = 1, \ldots, J \tag{6.3}$$

$$\sum_{k=1}^{K_i} \sum_{t=\mathrm{ES}_{ij}}^{\mathrm{LS}_{ij}} (t + d_{ijk}) x_{ijkt} = \sum_{k=1}^{K_i} \sum_{t=\mathrm{ES}_{i,j+1}}^{\mathrm{LS}_{i,j+1}} t x_{i,j+1,k,t}, \quad i \in W;\ j = 1, \ldots, J - 1 \tag{6.4}$$

$$\sum_{k=1}^{K_i} \sum_{t=\mathrm{ES}_{ij}}^{\mathrm{LS}_{ij}} (t + d_{ijk}) x_{ijkt} \le \sum_{k=1}^{K_i} \sum_{t=\mathrm{ES}_{i,j+1}}^{\mathrm{LS}_{i,j+1}} t x_{i,j+1,k,t}, \quad i \in \overline{W};\ j = 1, \ldots, J - 1 \tag{6.5}$$

$$\sum_{i=1}^{I} \sum_{j=1}^{J} \sum_{k=1}^{K_i} r_{ikl} \cdot \sum_{\tau=\max\{\mathrm{ES}_{ij}, t-d_{ijk}\}}^{\min\{t-1, \mathrm{LS}_{ij}\}} x_{ijk\tau} \le R_l, \quad l = 1, \ldots, L;\ t = 1, \ldots, T \tag{6.6}$$

$$x_{ijkt} = \{0, 1\}, \quad i = 1, \ldots, I; \; j = 1, \ldots, J; \; k = 1, \ldots, K_i; \; t = 1, \ldots, T \tag{6.7}$$

The objective function (6.1) aims to minimize project duration. It is equal to the finish time of sub-activity a_{IJ}. ES_{ij} and LS_{ii} represent the earliest and latest start times of a_{ii}, respectively. Suppose that all activities are performed by their fastest execution modes; then time parameters ES_{ij} and LS_{ii} can be gotten from the traditional forward and backward pass calculations. The backward pass recursion is started from a project length T, which equals a feasible project length. Binary variable x_{ijkt} is equal to 1 if sub-activity a_{ij} is performed in mode k and started at time t, and 0 otherwise. Constraints (6.2) ensure that each sub-activity is performed in exactly one mode. Constraints (6.3) take the finish to start precedence relations with a lag time of zero into account, where E_{FS} denotes the resulting set of finish to start precedence relations. The logical relations between different units of an activity are guaranteed by constraints (6.4) if this activity is required to maintain the resource continuity; otherwise, constraints (6.5) are adopted. Set W consists of those activities that must satisfy the resource continuity constraint; the other activities are classified into set $\overline{W}$. The renewable resource constraints are guaranteed by constraints (6.6), where T can be said to be an upper bound on the project duration. Constraints (6.7) force the decision variables to be binary values.

If other types of time constraints need to be considered, that is, start to start, start to finish, and finish to finish, then we just need to add the following corresponding constraints into the mode.

$$\sum_{k=1}^{K_i} \sum_{t=\mathrm{ES}_{ij}}^{\mathrm{LS}_{ij}} t x_{ijkt} \leq \sum_{k=1}^{K_s} \sum_{t=\mathrm{ES}_{sj}}^{\mathrm{LS}_{sj}} t x_{sjkt}, \quad (i,s) \in E_{\mathrm{SS}}; \; j = 1, \ldots, J \tag{6.8}$$

$$\sum_{k=1}^{K_i} \sum_{t=\mathrm{ES}_{ij}}^{\mathrm{LS}_{ij}} t x_{ijkt} \leq \sum_{k=1}^{K_s} \sum_{t=\mathrm{ES}_{sj}}^{\mathrm{LS}_{sj}} (t + d_{sjk}) x_{sjkt}, \quad (i,s) \in E_{\mathrm{SF}}; \; j = 1, \ldots, J \tag{6.9}$$

$$\sum_{k=1}^{K_i} \sum_{t=\mathrm{ES}_{ij}}^{\mathrm{LS}_{ij}} (t + d_{ijk}) x_{ijkt} \leq \sum_{k=1}^{K_s} \sum_{t=\mathrm{ES}_{sj}}^{\mathrm{LS}_{sj}} (t + d_{sjk}) x_{sjkt}, \quad (i,s) \in E_{\mathrm{FF}}; \; j = 1, \ldots, J \tag{6.10}$$

where E_{SS}, E_{SF}, and E_{FF} denote the resulting set of start to start, start to finish, and finish to finish precedence relations with zero lag time, respectively.

6.3 GA-BASED METHOD

The concept of genetic algorithms (GAs) originates from biology. Biologically, genes of a good parent produce better offspring. GAs search a problem space with a population of chromosomes and select chromosomes for continued search based on their performance. In GAs, potential solutions to a problem are represented as a population of chromosomes, and each chromosome stands for a possible solution at hand. The chromosomes evolve through successive generations. Offspring chromosomes are created by merging two parent chromosomes using a crossover operator, or modifying a chromosome using a mutation operator. During each generation, the chromosomes are evaluated on their performance with respect to the fitness functions. Fitter chromosomes have higher survival probabilities. After several generations, chromosomes in the new generation may be identical, or certain termination conditions are met. The final chromosomes hopefully represent optimal or near-optimal solutions to a problem. According to previous ideas, GAs contain two major ingredients: chromosome representation and genetic operators.

6.3.1 Chromosome Representation

For multi-mode resource-constrained scheduling in repetitive construction projects, the basic work of chromosome representation aims at deciding (1) the priority values of sub-activities that determine which sub-activity is scheduled first when resource conflicts occur, and (2) the modes to be selected for sub-activities. We use the mode list to show the execution modes of all sub-activities. The mode list is divided into I segments, of which the j ($j=1,\ldots,J$) gene in segment i ($i=1,\ldots,I$) is valued by one of the modes of activity i, symbolizing the modes to be selected for sub-activity a_{ij}. For the priority values, we adopt the priority list in the genetic representation introduced in Gen and Cheng (2000). For the mode list, the priority list is divided into I segments as well, of which the gene j in segment i represents the priority value of activity a_{ij}. The random value of each gene is an integer exclusively within $[1, I \times J]$ (I and J are the number of activities and units in a

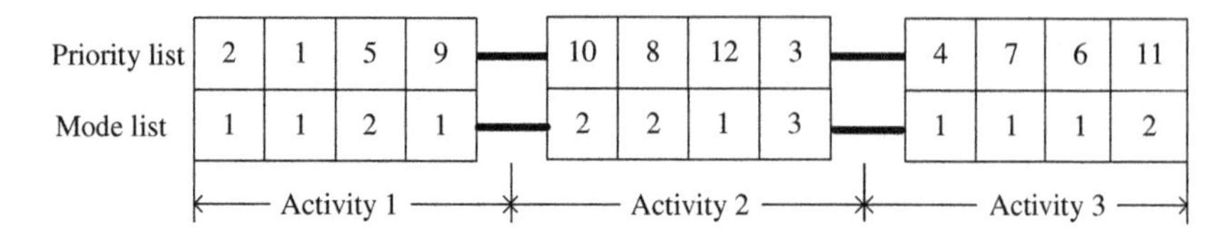

Figure 6.1 The structure of a chromosome.

project, respectively), and the larger the integer, the higher the priority. Figure 6.1 is an example of the structure of a chromosome.

In this chapter, the serial schedule scheme is used to generate project plans from the chromosome. In repetitive construction projects, not only the resource constraints, but the resource continuity constraint must be considered. Therefore the traditional serial schedule scheme cannot serve directly for resource-constrained scheduling in repetitive construction projects. We make some modifications to serial schedule scheme as follows. The algorithm divides a schedule procedure into several stages. Let PS_h denote the partial scheduled plan at stage h; E_{h+1} represents the feasible set of sub-activities that meet the precedence relations at stage $h+1$. Suppose that a_{ij} is a sub-activity with the highest priority value in E_{h+1}, and activity i is required to meet the resource continuity constraint. ES_{ij} is the earliest start time of a_{ij} and is valued by the maximum finish time of all the precedence sub-activities of a_{ij}. If ES_{ij} is greater than the finish time of $a_{i,j-1}$, both the sub-activities a_{in}, $n = 1, 2, \ldots, j-1$, need to be delayed until the resource continuity constraint is met. Then we check the renewable resource constraint for this sub-activity. Repeat the above operations until all constraints are met. When the feasible project plan is decided, the reciprocal of project duration is taken as the fitness of the corresponding chromosome.

6.3.2 Selection Operator

We consider several variants of the selection operator, all of which comply with a survival of the fittest strategy. Let POP denote the size of the population; the roulette wheel selection calculates the selection probability of individuals and selects the best ones; the others are deleted from the population. Then, in the tournament selection, a number of individuals compete for survival. These competitions, in which the least fit individual is removed from the population, are repeated until POP individuals are left. Finally, we consider that the selection operator consists of roulette wheel selection and preservation of the best individual; the selection probability of individual I, denoted by $p(I)$, is formulated as

$$p(I) = \frac{f(I)}{\sum_{x=1}^{POP} f(x)} \tag{6.11}$$

where $f(I)$ is the fitness value of individual I.

6.3.3 Crossover Operator

POP pairs of individuals are randomly selected by the above operation to undergo the crossover operation. For the mode list, the one point crossover operator is used. A random integer p_{cross} with $1 \leq p_{cross} \leq I \times J$ is generated as a crossover point. The positions $i = 1, \ldots, p_{cross}$ in the child are taken for the genes from one parent; that is, $m_c(i) = m_f(i)$, where $m_c(i)$ is the value of the position i in the child and $m_f(i)$ is the value of the position i in the father. Meanwhile, the remaining positions $i = p_{cross} + 1, \ldots, I \times J$ in the child are filled with genes from the other parent; that is, $m_c(i) = m_m(i)$, where $m_m(i)$ is the value of the position i in the mother. For the priority list, the position-based crossover operator proposed by Syswerda (1991) is adopted. Essentially, the child takes some genes from one parent randomly and fills vacuum positions with genes from the other parent by a left-to-right scan. Figure 6.2 illustrates the crossover operation with an example.

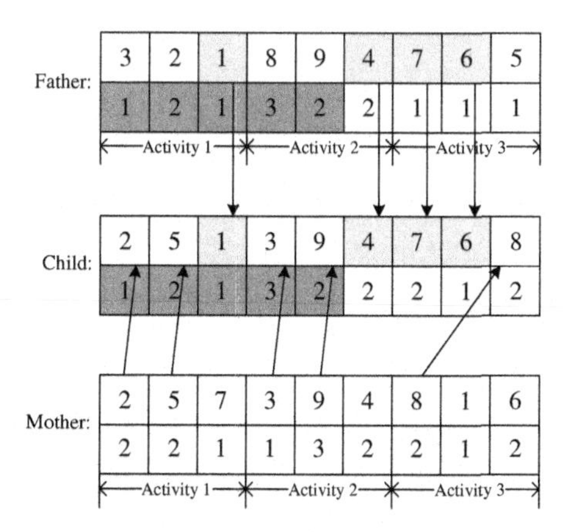

Figure 6.2 Example of crossover operation.

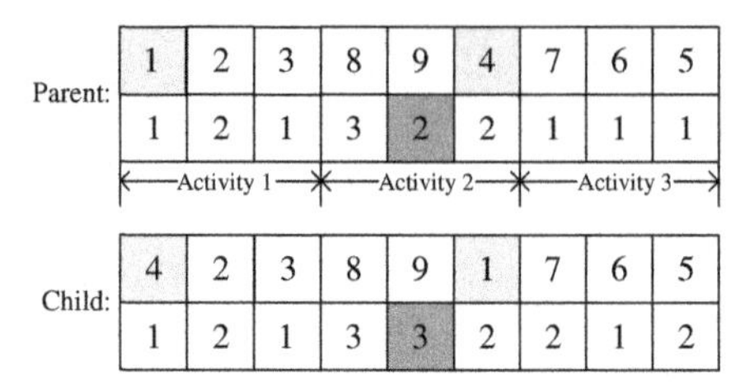

Figure 6.3 Example of mutation operation.

6.3.4 Mutation Operator

The mutation operator is applied on newly generated individuals with a probability of mutation p_{mut}. This operator is applied, first, on the priority list string and, second, in the mode assignment string. In the first substage, the mutation operator proposed by Elloumi and Fortemps (2010) is utilized. We randomly choose two positions q_{mut1} and q_{mut2} such that $q_{\text{mut1}} < q_{\text{mut2}}$; then we check whether jobs in positions q_{mut1} and q_{mut2} can be permutated; that is, if the job in position q_{mut2} is not an immediate successor of the job in position q_{mut1}, we can permutate the jobs; otherwise, we randomly choose the other two positions q'_{mut1} and q'_{mut2} with $q'_{\text{mut1}} < q'_{\text{mut2}}$. We repeat the procedure until two jobs are permutated or until J unsuccessful operations are made. Note that at this stage, the mode assignment is not affected; that is, permutated jobs keep their initially assigned modes. Afterwards, we randomly select one job q_{mut3} which has more than one mode alternative. From the set of modes of this activity, we randomly assign a mode different from the current one. Figure 6.3 is an example of mutation operation.

6.4 CASE STUDY

Take the concrete bridge construction project presented first by Selinger (1980) as an example (Figure 6.4). It is a serial project that consists of four units, each of which includes the following activities in sequence: excavation, foundation, Columns, beams, and slabs. Each activity is to be performed by a single crew advancing from one unit to the next in the given order of units 1 to 4. The precedence relations among activities are finish to start with zero lag time. The basic data for the project, including the labor requirement of each activity, the feasible quantities of resources (i.e., the alternative modes), and the

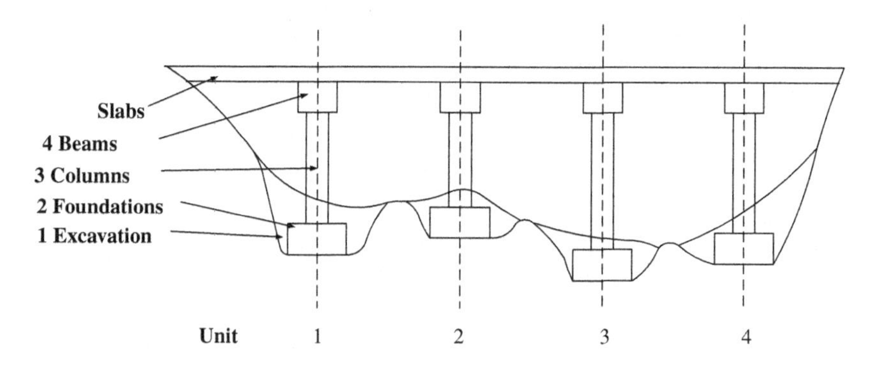

Figure 6.4 Example project.

Table 6.1 Project Information

Activity	Labor Requirement in Hours				Feasible Quantities of Workers				Daily Working Hours
	Unit 1	Unit 2	Unit 3	Unit 4	Mode 1	Mode 2	Mode 3	Mode 4	
1. Excavation	600	750	520	800	6				8
2. Foundation	920	960	840	800	10	8	6		8
3. Columns	1450	1200	1800	1400	10	12	14		8
4. Beams	480	520	570	450	7	6	5	4	8
5. Slabs	0	1140	940	1200	9	8			8

number of daily working hours for each activity, are given in Table 6.1. The duration of sub-activity a_{ij} in mode k, denoted by d_{ijk}, is equal to the labor requirement of a_{ij} divided by the product of the quantities of resources of activity i in mode k and the number of daily working hours of this activity.

Suppose that not all activities are required to meet the resource continuity constraint. When the maximum availability of workers per day is 15, the optimal repetitive scheduling method (RSM) schedule for the example project is shown in Figure 6.5, the optimal start and finish times of all sub-activities are listed in Table 6.2, and the optimal execution modes for all sub-activities are presented in Table 6.3. From the results obtained, the minimum duration of the project is 170.56 days.

If the limitation for resources is not involved, the shortest project duration can be calculated using the optimization algorithm presented in Chapter 5. Since the precedence relations among activities are finish to start and all activities are not required to meet the resource

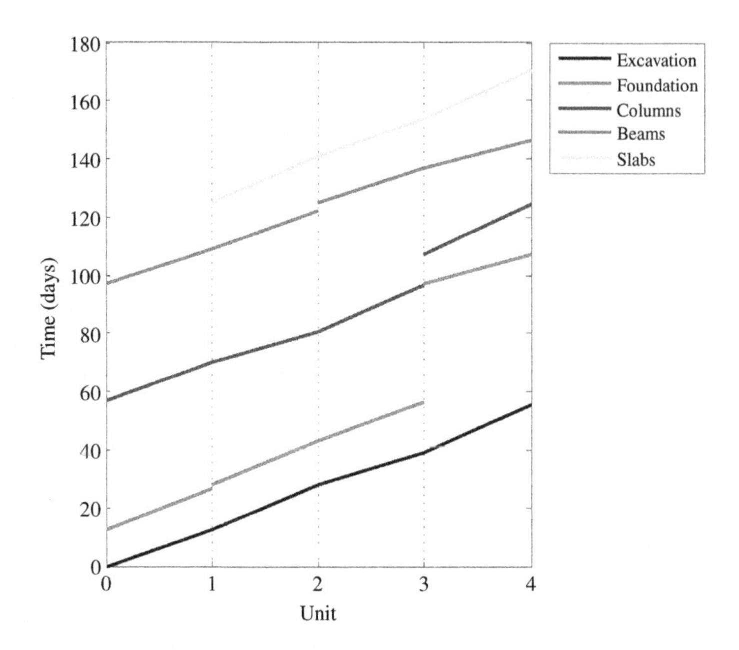

Figure 6.5 Optimal RSM schedule for the example project (maximum resource availability per day: 15).

Table 6.2 Optimal Time Parameters of All Sub-Activities (Maximum Availability of Workers per Day: 15)

Activity		1		2		3		4		5	
Timings		Start	Finish	Start	Finish	Start	Finish	Start	Finish	Start	Finish
Unit	1	0	12.50	12.50	26.88	57.00	69.95	97.00	109.00	–	–
	2	12.50	28.13	28.13	43.13	69.95	80.67	109.00	122.00	125.00	140.83
	3	28.13	38.96	43.13	56.25	80.67	96.73	125.00	136.88	140.83	153.89
	4	38.96	55.63	97.00	107.00	107.00	124.50	136.88	146.25	153.89	170.56

Table 6.3 Optimal Execution Modes of All Sub-Activities (Maximal Availability of Workers per Day: 15)

Activity		1		2		3		4		5	
Mode		No.	Resource Consumption	No.	Resource Consumption	No.	Resource Consumption	No.	Resource Consumption	No.	Resource Consumption
Unit	1	1	6	2	8	3	14	3	5	–	–
	2	1	6	2	8	3	14	3	5	1	9
	3	1	6	2	8	3	14	2	6	1	9
	4	1	6	1	10	1	10	2	6	1	9

continuity constraint, there is no activity that meets the prerequisites for a segment to be a backward controlling segment (see Theorem 5.1). Therefore, the optimal solution can be obtained directly by performing all sub-activities in their fastest execution modes and starting all sub-activities at their earliest times. The optimal RSM schedule without resource limitation is shown in Figure 6.6, where the shortest project duration is reduced to 106.81 days.

As shown in Figure 6.7, when the maximal resource availability of workers per day increases from 14 to 22, the shortest project duration increases from 130.7 days to 171.6 days.

6.4.1 The Effect of Resource Continuity on Project Scheduling

One of the characteristics of repetitive project scheduling is that it should keep the continuity of resources for some activities strictly when needed. Therefore, the effect of resource continuity on project scheduling needs to be analyzed further. Now we introduce a new constraint into the above example: activities "Columns" and "Beams" must be performed without interruption. Assume that the maximum availability of workers per day is equal to 15; then the shortest project duration calculated by the proposed GA is 176.6 days, an increase of 6 days compared with the case where the resource continuity constraint is not required for all activities. In this case, the optimal RSM schedule calculated by the proposed GA is shown in Figure 6.8, the optimal time parameters of all sub-activities are listed in Table 6.4, and the optimal execution modes of all sub-activities are presented in Table 6.5.

6.4.2 Comparison of Different Algorithms

It is the evolutionary algorithm proposed by Hsie et al. (2009) that needs to be compared. It should be noted that this algorithm assumes that all activities can reselect their execution modes in different time periods, rather than different units in space segments. Because Hsie et al. (2009) required all activities to meet the resource continuity constraint, we adopt the same assumption for comparison purposes. Using one day as the minimum time period for the Hsie et al. algorithm, the comparative results of the two algorithms are shown in Table 6.6, where the maximum resource availability per day progressively increases from 13 to 21.

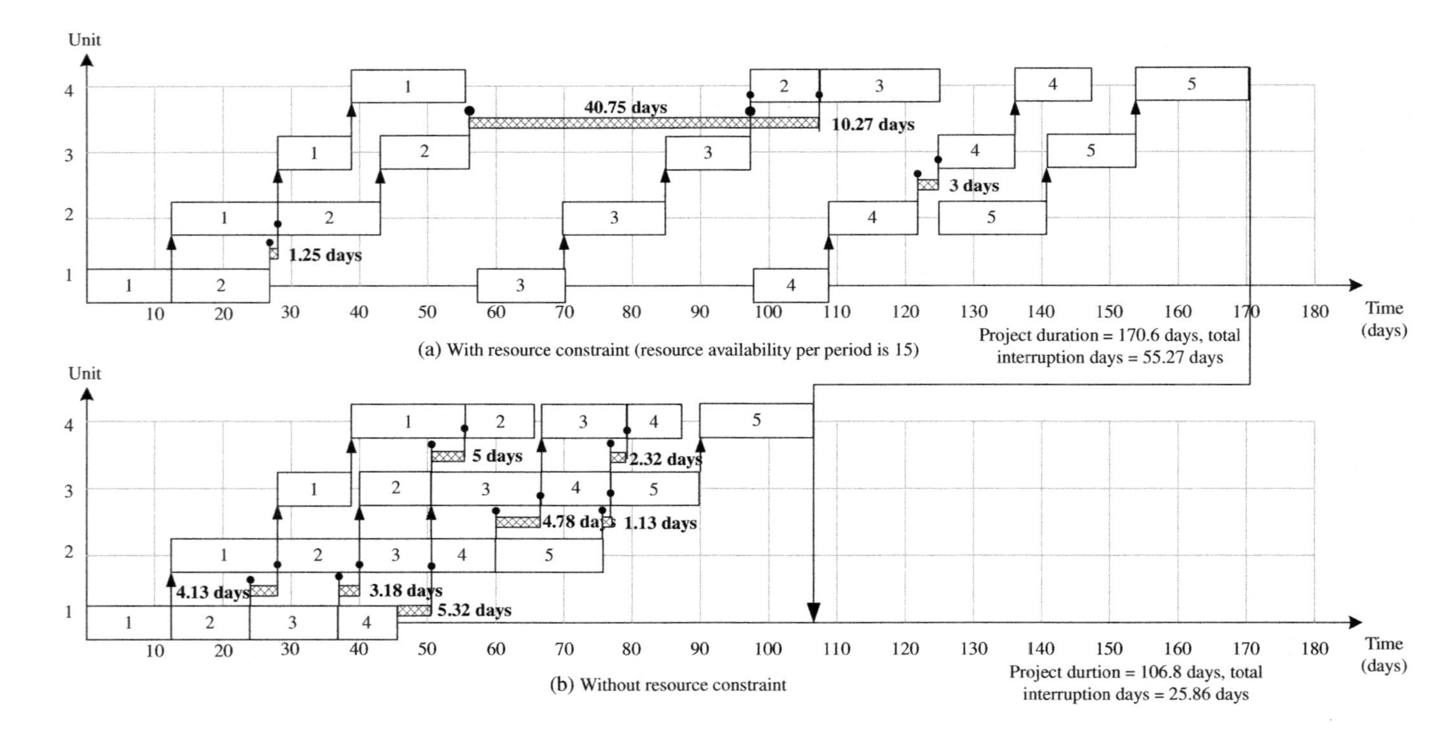

Figure 6.6 Comparison between the optimal schedule with and without resource constraint.

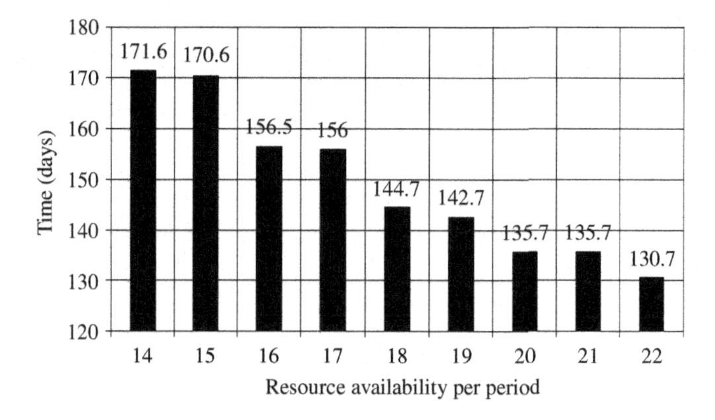

Figure 6.7 Relationship between maximal resource availability per day and project duration.

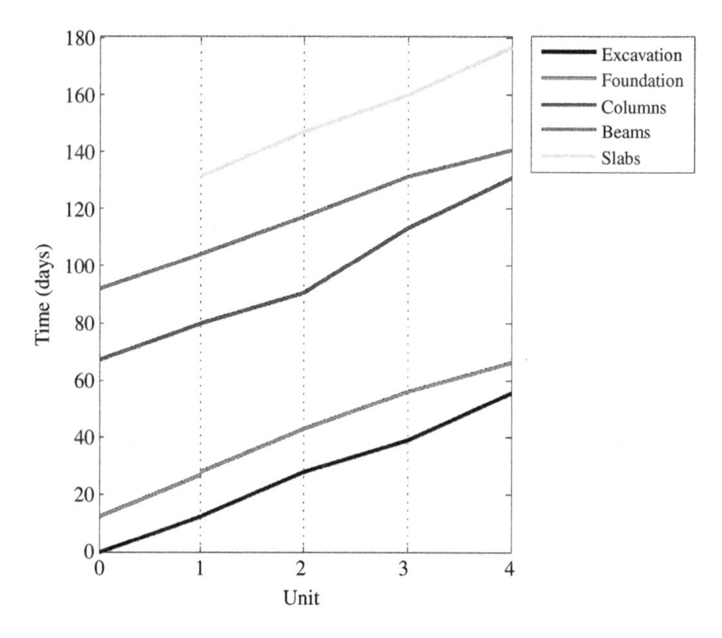

Figure 6.8 Optimal RSM schedule for the example project (maximum resource availability per day: 15/resource continuity constraint is required for activities 3 and 4).

For this example project, the project duration calculated by the proposed algorithm is always less than or equal to that calculated by Hsie et al. algorithm as the maximal availability of workers per day increases. In other words, it is not always good to require activities to maintain the same production rate in each time period.

Table 6.4 Optimal Time Parameters of All Sub-Activities (Maximum Resource Availability per Day: 15/ Resource Continuity Constraint Is Required for Activities 3 and 4)

Activity		1		2		3		4		5	
Timings		Start	Finish	Start	Finish	Start	Finish	Start	Finish	Start	Finish
Unit	1	0	12.50	12.50	26.88	67.00	79.95	92.00	104.00	–	–
	2	12.50	28.13	28.13	43.13	79.95	90.67	104.00	117.00	131.00	146.83
	3	28.13	38.96	43.13	56.25	90.67	113.16	117.00	131.25	146.83	159.89
	4	38.96	55.63	56.25	66.25	113.16	130.67	131.25	140.63	159.89	176.56

Table 6.5 Optimal Excavation Modes of All Sub-Activities (Maximum Resource Availability per Period: 15/Resource Continuity Constraint Is Required for Activities 3 and 4)

Activity		1		2		3		4		5	
Mode		No.	Resource Consumption	No.	Resource Consumption	No.	Resource Consumption	No.	Resource Consumption	No.	Resource Consumption
Unit	1	1	6	2	8	3	14	3	5	—	—
	2	1	6	2	8	3	14	3	5	1	9
	3	1	6	2	8	1	10	3	5	1	9
	4	1	6	1	10	1	10	2	6	1	9

Table 6.6 Comparison Results of the Two Algorithms

Maximum Resource Availability per Day	Project Duration (days)	
	Proposed Algorithm	Hsie et al. Algorithm
13	190.5	202.6
14	174.6	186.6
15	174.6	175.6
16	163.6	169.6
17	163.6	164.6
18	157.6	157.6
19	157.6	157.6
20	154.6	154.6
21	154.6	154.6

6.5 CONCLUSION AND PROSPECT

This chapter proposed a GA-based optimization model for the resource-constrained scheduling problem in repetitive construction projects. Unlike existing models, the proposed algorithm considers both multiple modes and work interruption, and aims at minimizing project duration by determining the optimum execution modes and start times for all sub-activities, while satisfying a system of precedence relation constraints, renewable resource constraints, and resource continuity constraints. The significance of this research lies in providing a more reasonable and better solution for resource-constrained scheduling in repetitive construction projects by allowing activities to be interrupted. An area for future research could be the study of integrating soft logic in resource-constrained scheduling problems in repetitive construction projects. By allowing activities to be performed in various work sequences, the planner can receive a more valuable and practical project schedule. This is of great importance in dealing with real-world applications. On the other hand, considering other objective functions such as net present value maximization and resource availability cost minimization would also be suitable future directions.

CHAPTER 7

Discrete Time–Cost Trade-Off in Repetitive Construction Projects

7.1 INTRODUCTION

Time–cost trade-offs have attracted growing attention in the construction industry for the purpose of time management and cost control in construction projects. Time–cost behavior in an activity describes the non-increasing functional relation between the duration of an activity and the amount of nonrenewable resources (e.g., money) committed to it. In the discrete version of the time–cost trade-off problem (DTCTP), the objective has usually been divided into three parts: (1) to minimize project cost without exceeding an allowed deadline (deadline problem); (2) to find the shortest project duration while meeting a given budget (budget problem); and (3) to construct the complete and efficient time–cost profile over the set of feasible project durations (time–cost curve problem). The DTCTP has been studied extensively under various assumptions in the late 1950s and in this chapter; we focus on this well-known problem in repetitive construction projects.

Scheduling decisions for repetitive construction projects are complex, since several cost elements related to different aspects of the project must be considered and balanced by the planner in order to construct a cost-efficient schedule (Ipsilandis, 2007). In repetitive construction projects, the total cost function often has three components: direct costs (resulting from the performance of project activities), indirect costs (those items that are not directly related to individual project activities), and idle resource costs (incurred by contactors during scheduled interruptions of the selected crew to cover the costs of idle resources).

Reda (1990) presented a linear programming formulation. The goal is to complete the project within the pre-specified target duration at a minimum direct cost; the constraints include maintaining production rates and continuity of work. Senouci and Eldin (1996) proposed a dynamic programming model to determine time/cost profiles of non-serial repetitive construction projects. Their model considered the

Repetitive Project Scheduling: Theory and Methods.

impacts of crew formation, interruptions, and lags for production activities, where the durations and interruptions of activities were described by continuous and discrete functions.

Hyari et al. (2009) adopted an approach based on a genetic algorithm (GA) and a Pareto ranking method to determine the Pareto front for the DTCTP in repetitive construction projects. Other time-related costs are also considered in the model, including early completion incentives, late completion penalties, and lane rental costs. Ezeldin and Soliman (2009) proposed a hybrid technique that combines GAs with dynamic programming to resolve the DTCTP in non-serial repetitive construction projects under uncertainty. Long and Ohsato (2009) presented an interesting model based on GA that aimed to minimize both project duration and total cost corresponding to different levels of the relative importance of these two objectives. The authors also identified two types of activities: type α activities that must be performed continuously to maintain work continuity, and type β activities that allow violations of the work continuity constraint. Terry and Lucko (2012) presented a new method for performing time–cost trade-off analysis in repetitive construction projects using the singularity function. This method established a comprehensive model that integrated the time and cost aspects by expressing their complex interaction with a single, versatile, and extendable singularity function.

The above studies differ from optimization methods, objective functions, and assumptions on resource continuity constraint. In spite of major differences, they consider two same hypotheses: the mode identity assumption and the hard logic assumption. In the repetitive scheduling method, the mode identity assumption requires that sub-activities of any activity on different units must be performed by the same execution mode, and the hard logic assumption requires that there is only one logical sequence of activities in a project. In practice, however, changing the execution mode for different units is allowed and is prevalent in real projects, and is considered an effective way of controlling project cost. Moreover, the work sequence among units of an activity is not necessarily unchangeable in many repetitive construction projects. In other words, the activities may be of a "soft" character. Current optimization methods do not account for this fact and, therefore, require a planner to arbitrarily determine a particular logical sequence based on some assumptions. In the worst case, the sequence

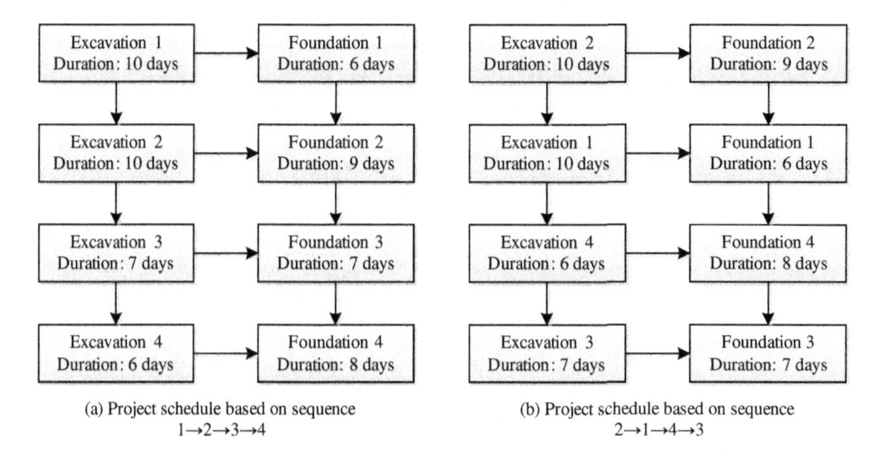

Figure 7.1 An example of soft logic.

chosen by the planner may be far from the optimal schedule in terms of time and cost. One such example is shown in Figure 7.1. Previous research assumed that the work of "excavation" and "foundation" could be performed only in the sequence 1→2→3→4; the resulting project duration is 44 days. However, if the logical sequence of both activities is set to 2→1→4→3, the project duration reduces to 41 days. Consequently, the overall project cost will also decrease due to the decrease in the indirect project costs.

In this chapter, two types of new DTCTPs in repetitive construction projects are considered. The first is mode-variable DTCTP and the second is the DTCTP with soft logic. The mathematical models and optimization methods of these two problems are described in Sections 7.2 and 7.3, respectively.

7.2 MODE-VARIABLE DISCRETE TIME–COST TRADE-OFF PROBLEM (MVDTCTP)

7.2.1 Problem Formulation

We consider a repetitive construction project that consists of I activities and each activity is repeated in J units. Activity 1 is the only start activity and activity I is the only finish activity. Activity i, $i = 1,\ldots, I$, is connected with its preceding activity p by fulfilling the precedence relationship of finish to start with zero lag time; $p \in P_i$, where P_i is the set of predecessors of activity i. Sub-activity a_{ij} may be executed in one of K_i modes. Let D_{ijk} denote the duration of a_{ij} performed by mode k.

Project cost is measured as the sum of direct costs, idle resource costs in all units of all activities, and the project indirect cost. The direct costs for each sub-activity a_{ij} include labor cost, material cost, and equipment cost. Specifically, the labor cost of a_{ij} performed by mode k, denoted by LC_{ijk}, is calculated using the formula $\mathrm{LC}_{ijk} = D_{ijk}L_{ik}$, where L_{ik} is the daily labor cost of activity i in mode k. Similarly, the equipment cost of a_{ij} performed by mode k, denoted by EC_{ijk}, is calculated using the formula $\mathrm{EC}_{ijk} = D_{ijk}E_{ik}$, where E_{ik} is the daily equipment cost of activity i in mode k. MC_{ij} denotes the material cost of a_{ij} and is calculated using the formula $\mathrm{MC}_{ij} = Q_{ij}M_i$, where Q_{ij} is the quantity of materials required to complete activity i in unit j and M_i is the corresponding unit cost of these materials.

Idle resource costs are incurred by contractors during scheduled interruptions of the selected mode to cover the costs of idle resources. In this chapter, the idle resource cost of each activity is measured by the product of the maximum daily labor cost of all selected modes in the activity and the corresponding interruption days. Indirect project cost, denoted by IC, increases linearly with the project duration; it is calculated by the formula $\mathrm{IC} = F_t\mathrm{ICR}$, where ICR is the daily indirect cost of the project and F_t denotes the project duration.

This chapter introduces a binary variable x_{ijk} to determine the execution mode of each sub-activity a_{ij}; $x_{ijk} = 1$ if a_{ij} is performed by mode k, and 0 otherwise. Variables S_{ij} and INT_{ij} denote the start time of a_{ij} and the time for which resources remain idle after a_{ij} is completed. The objective is to find the minimal project cost while meeting a given deadline. Finally, the mode-variable discrete time–cost trade-off problem (MVDTCTP) can be mathematically described as follows:

$$\min\ \mathrm{TC} = \sum_{i=1}^{I}\sum_{j=1}^{J}\sum_{k=1}^{K_i}\{x_{ijk}(\mathrm{LC}_{ijk} + \mathrm{EC}_{ijk}) + \mathrm{MC}_{ij} + \mathrm{IRC}_i\} + F_t\mathrm{ICR} \tag{7.1}$$

$$S_{pj} + \sum_{k=1}^{K_p} x_{pjk}D_{pjk} \le S_{ij}, \quad p \in P_i,\ i = 1,\ldots,I,\ j = 1,\ldots,J \tag{7.2}$$

$$S_{ij} + \sum_{k=1}^{K_i} x_{ijk}D_{ijk} + \mathrm{INT}_{ij} - S_{i,j+1} = 0, \quad i = 1,\ldots,I,\ j = 1,\ldots,J-1 \tag{7.3}$$

$$\sum_{j=1}^{J} x_{ijk} L_{ik} \left(\sum_{l=1}^{J-1} \mathrm{INT}_{il} \right) \leq \mathrm{IRC}_i, \quad i=1,\ldots,I, \; k=1,\ldots,K_i \tag{7.4}$$

$$S_{IJ} + \sum_{k=1}^{K_I} x_{IJk} D_{IJk} \leq F_t \tag{7.5}$$

$$F_t \leq T_{\max} \tag{7.6}$$

$$\sum_{k=1}^{K_i} x_{ijk} = 1, \quad i=1,\ldots,I, \; j=1,\ldots,J \tag{7.7}$$

The objective function (7.1) minimizes the project cost TC. Constraints (7.2) guarantee that the FS precedence relations between activities in each unit are preserved. Constraints (7.3) ensure that every activity is performed according to the unique logical sequence of units 1 to J. Constraints (7.4) estimate the values of IRC_i for all $i=1,\ldots,I$. Constraint (7.5) ensures that the project duration is greater than the finish times of the finish sub-activity a_{IJ}. Constraint (7.6) forces the project duration to not exceed the given deadline. Constraints (7.7) require every sub-activity to be performed in only one mode.

7.2.2 Available Optimization Method

In repetitive construction projects, the difference between MVDTCTP and DTCTP lies only in the selection of execution modes of sub-activities. That is, the MVDTCTP allows different units of an activity to be executed by different modes, while the DTCTP requires that all units of an activity must be performed by the same mode. Therefore, the existing optimization methods for the DTCTP can be modified to solve the MVDTCTP. The GA presented by Hyari et al. (2009) is used with the following modifications: (1) in the encoding operator, each gene denotes the execution mode of a particular sub-activity rather than an activity. Then the length of the mode list should be increased to $I \times J$, where I and J denote the number of activities and units of a project, respectively. (2) In the fitness calculation, the idle resource cost of each activity is measured by the product of the maximum daily labor cost of all selected modes in the activity and the corresponding interruption days.

7.2.3 Case Study

A concrete bridge project that was first presented by Selinger (1980) is analyzed. This project consists of four units, and each unit involves the following five activities in sequence: excavation, foundation, columns, beams, and slabs. Table 7.1 presents project data on the quantities of work for the five activities in each of the four units and the available modes along with their productivity rates and daily costs. Based on Table 7.1, D_{ijk} can be calculated using the formula $D_{ijk} = Q_{ij}/P_{ik}$. We assume that the indirect project cost per day is $2500, in line with related literature.

Table 7.2 lists all non-dominated solutions for the project analyzed, where it is not possible to find another solution that provides lower project duration and lower project direct cost at time same time. The minimum project direct cost is $1,317,642 and the corresponding project duration is 143 days. The shortest project duration was 123 days and the corresponding project direct cost was $1,654,032.

Table 7.1 Tabular Presentation of Project Data

Activity (i)	**Excavation (i = 1)**				**Foundation (i = 2)**				**Columns (i = 3)**			
Unit (j)	**1**	**2**	**3**	**4**	**1**	**2**	**3**	**4**	**1**	**2**	**3**	**4**
Quantity of work ($Q_{i,j}$) in m^3	1147	1434	994	1529	1032	1077	943	898	104	86	129	100
Mode (k)	1				1	2	3		1	2	3	
Productivity (P_{ik}) in m^3/day	91.75				89.77	71.81	53.86		5.73	6.88	8.03	
Labor cost in $/day	340				3804	2853	1902		1875	2438	3000	
Equipment cost in $/day	566				874	655	436		285	371	456	
Material cost in $/m^3	0				92				479			

Activity (i)	**Beams (i = 4)**				**Slabs (i = 5)**			
Unit (j)	**1**	**2**	**3**	**4**	**1**	**2**	**3**	**4**
Quantity of work ($Q_{i,j}$) in m^3	85	92	101	80	0	138	114	145
Mode (k)	1	2	3	4	1	2		
Productivity (P_{ik}) in m^3/day	9.9	8.49	7.07	5.66	8.73	7.76		
Labor cost in $/day	3931	3238	2544	1850	2230	1878		
Equipment cost in $/day	315	259	204	148	177	149		
Material cost in $/m^3	195				186			

Table 7.2 Non-Dominated Solutions Generated by the MVDTCTP Model

No.	Project Duration (days)	Direct Cost (dollars)	Indirect Cost (dollars)	Total Cost (dollars)	Execution Mode
1	107	1,448,851	267,500	1,716,351	{1 1 1 1},{2 1 1 1},{3 3 3 3},{3 1 1 3},{ − 1 1 1}
2	108	1,431,152	270,000	1,701,152	{1 1 1 1},{2 1 1 1},{3 3 3 3},{3 2 2 3},{ − 1 1 1}
3	109	1,428,208	272,500	1,700,708	{1 1 1 1},{2 1 1 1},{3 3 3 3},{2 3 2 3},{ − 1 1 1}
4	110	1,422,183	275,000	1,697,183	{1 1 1 1},{2 1 1 1},{3 3 3 3},{3 3 2 4},{ − 1 1 1}
5	111	1,407,822	277,500	1,685,322	{1 1 1 1},{2 2 1 2},{2 3 3 3},{3 2 2 3},{ − 1 1 1}
6	112	1,403,933	280,000	1,683,933	{1 1 1 1},{2 1 2 2},{3 2 3 3},{4 2 2 3},{ − 1 1 1}
7	113	1,394,200	282,500	1,676,700	{1 1 1 1},{2 1 2 2},{3 2 3 2},{3 3 3 3},{ − 1 1 1}
8	114	1,387,152	285,000	1,672,152	{1 1 1 1},{2 1 2 2},{3 2 3 3},{4 3 3 4},{ − 1 1 1}
9	115	1,383,406	287,500	1,670,906	{1 1 1 1},{2 1 2 3},{3 2 3 2},{4 3 3 4},{ − 1 1 1}
10	116	1,376,636	290,000	1,666,636	{1 1 1 1},{2 2 2 3},{2 2 3 2},{3 3 3 3},{ − 1 1 1}
11	117	1,368,040	292,500	1,660,540	{1 1 1 1},{2 2 2 3},{2 2 3 3},{4 3 3 4},{ − 1 1 1}
12	118	1,364,006	295,000	1,659,006	{1 1 1 1},{2 2 2 3},{2 2 3 2},{4 3 3 4},{ − 1 1 1}
13	119	1,361,700	297,500	1,659,200	{1 1 1 1},{2 2 2 3},{2 1 3 2},{4 3 3 4},{ − 1 1 1}
14	120	1,355,274	300,000	1,655,274	{1 1 1 1},{2 1 3 3},{2 1 3 2},{4 4 3 4},{ − 1 1 1}
15	121	1,354,658	302,500	1,657,158	{1 1 1 1},{2 1 3 3},{3 1 3 1},{4 4 4 3},{ − 2 1 1}
16	122	1,349,391	305,000	1,654,391	{1 1 1 1},{2 1 3 3},{2 1 3 1},{4 4 4 3},{ − 2 1 1}
17	123	1,346,532	307,500	1,654,032	{1 1 1 1},{2 1 3 3},{2 1 3 1},{4 4 4 4},{ − 2 1 1}
18	124	1,344,883	310,000	1,654,883	{1 1 1 1},{2 1 3 3},{2 1 3 1},{4 4 4 4},{ − 2 2 1}
19	125	1,342,622	312,500	1,655,122	{1 1 1 1},{2 2 3 3},{1 1 3 1},{4 4 4 3},{ − 2 1 1}
20	126	1,339,768	315,000	1,654,768	{1 1 1 1},{2 2 3 3},{1 1 3 1},{4 4 4 4},{ − 2 1 1}
21	127	1,338,114	317,500	1,655,614	{1 1 1 1},{2 2 3 3},{1 1 3 1},{4 4 4 4},{ − 2 2 1}
22	128	1,337,665	320,000	1,657,665	{1 1 1 1},{2 2 3 3},{1 1 3 1},{4 4 4 4},{ − 2 1 2}
23	129	1,335,263	322,500	1,657,763	{1 1 1 1},{2 2 3 3},{1 1 2 1},{4 4 4 4},{ − 2 2 1}
24	130	1,332,498	325,000	1,657,498	{1 1 1 1},{3 2 3 3},{1 1 3 1},{4 4 4 4},{ − 2 2 1}
25	132	1,330,395	330,000	1,660,395	{1 1 1 1},{3 2 3 3},{1 1 3 1},{4 4 4 4},{ − 2 2 2}
26	133	1,329,647	332,500	1,662,147	{1 1 1 1},{3 2 3 3},{1 1 2 1},{4 4 4 4},{ − 2 2 1}
27	134	1,329,197	335,000	1,664,197	{1 1 1 1},{3 2 3 3},{1 1 2 1},{4 4 4 4},{ − 2 1 2}
28	135	1,326,637	337,500	1,664,137	{1 1 1 1},{3 3 3 3},{1 1 3 1},{4 4 4 4},{ − 2 2 1}
29	136	1,325,607	340,000	1,665,607	{1 1 1 1},{3 2 3 3},{1 1 1 1},{4 4 4 4},{ − 2 2 1}
30	137	1,323,787	342,500	1,666,287	{1 1 1 1},{3 3 3 3},{1 1 2 1},{4 4 4 4},{ − 2 2 1}
31	139	1,321,682	347,500	1,669,182	{1 1 1 1},{3 3 3 3},{1 1 2 1},{4 4 4 4},{ − 2 2 2}
32	140	1,321,397	350,000	1,671,397	{1 1 1 1},{3 3 3 3},{1 1 1 1},{4 4 4 4},{ − 2 1 1}
33	141	1,319,745	352,500	1,672,245	{1 1 1 1},{3 3 3 3},{1 1 1 1},{4 4 4 4},{ − 2 2 1}
34	142	1,319,296	355,000	1,674,296	{1 1 1 1},{3 3 3 3},{1 1 1 1},{4 4 4 4},{ − 2 1 2}
35	143	1,317,642	357,500	1,675,142	{1 1 1 1},{3 3 3 3},{1 1 1 1},{4 4 4 4},{ − 2 2 2}

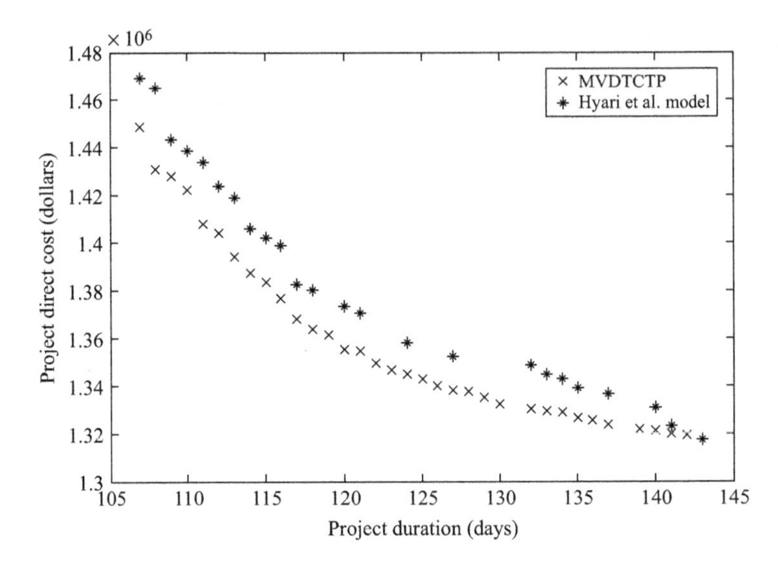

Figure 7.2 Calculation results of different models.

For this example, Hyari et al. (2009) have given the complete time–cost profile based on the mode identity assumption. Figure 7.2 compares this profile with the results calculated by the MVDTCTP model. Obviously, the MVDTCTP model can always get a more optimal solution than the Hyari et al. method.

7.3 THE DISCRETE TIME–COST TRADE-OFF PROBLEM WITH SOFT LOGIC (DTCTP-*SL*)

7.3.1 Soft Logic Concept and Its Applications in Project Management

Soft logic refers to those relations which allow connected activities to be scheduled by a variety of logical sequences, or performed simultaneously in certain circumstances. If the logical relation of an activity is soft, the planner can consider employing more resources to run the activities in parallel. As for repetitive activities, more investment in resources means a larger number of units that can be done concurrently (NUDC). As illustrated in Figure 7.3, if the NUDC of the activity "Excavation" is equal to 2, "Excavation 1" and"Excavation 2" can proceed simultaneously. Then "Excavation 3" will begin upon the completion of "Excavation 1," and "Excavation 4" will begin upon the completion of "Excavation 2," as shown in Figure 7.3(a). If we continue to increase the supply of resources (i.e., the NUDC changes to 3 or 4), "Excavation 3" or even

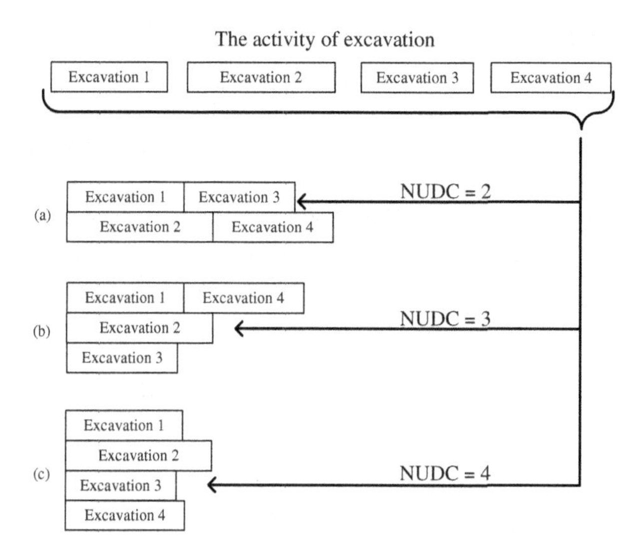

Figure 7.3 Example of soft relation in repetitive activity.

"Excavation 4" can be performed in parallel with "Excavation 1," as shown in Figure 7.3(b) and (c), respectively.

Tamimi and Diekmann (1988) developed the SOFTCPM method, which allowed for soft relations to update network models in cases where there was a possibility to change activity sequence. El-Sersy (1992) adopted some of the terminology associated with GERT to develop three types of soft links: OR (which allows running activities in parallel), EXCLUSIVE-OR (which allows reversing sequence), and SOFT (which allows canceling the relation). Wang (2005) examined the impact of soft relations on the duration of construction projects in stochastic conditions. The research reveals that the results are close to those obtained with PERT. Nevertheless, he assumed that the original predecessors and successors of activities do not change when soft links are ignored. Fan and Tserng (2006), utilizing the soft logic sequencing principles used in OERT, developed a computer system which provides the shortest duration logic and start and finish dates required to maintain work continuity in repetitive projects. Recently, Fan et al. (2012) presented a GA-based optimization model to search for the optimum activity mode and logical sequence that yield the minimum project cost. However, idle resource cost was not considered, so the model does not guarantee that the optimal schedule (in terms of costs) will be found.

Using soft logic can provide more flexibility in reducing project cost, activity timing, and resource allocation. In global market competition, both speed and efficiency as the competitive factors are gaining increasing importance for companies. Therefore, in order to arrive at a more competitive schedule, it is necessary to take soft logic into consideration when dealing with the DTCTP in repetitive construction projects. But this throws down a greater challenge for the planner to find the optimum sequence for the activities. To be specific, consider a project which consists of N activities, and each activity is repeated in M units. The number of solutions where soft logic is involved is $(M!)^N$ times that based on the assumption of fixed activity sequence.

7.3.2 Description of DTCTP-*sl*

Considering soft logic, the work sequence among units of each activity can be changed. For simplicity, we assume that no more than one unit of an activity can be performed concurrently in consideration of the limited resources. In other words, only one crew is available for each activity. Moreover, the mode identity assumption is adopted for all activities; that is, all units of an activity should be performed in the same mode. Binary variable $x_{ik}=1$, if the execution mode k is selected for activity I; otherwise, $x_{ik}=0$. Binary variable $y_{ijl}=1$ if sub-activity a_{ij} is scheduled in sequence l, and 0 otherwise. The objective is to find the minimal project cost while meeting a given deadline. Finally, the model of DTCTP-*sl* can be described as follows:

$$\min\ \mathrm{TC}=\sum_{i=1}^{I}\sum_{j=1}^{J}\sum_{k=1}^{K_i}\{x_{ik}(\mathrm{LC}_{ijk}+\mathrm{EC}_{ijk})+\mathrm{MC}_{ij}+\mathrm{IRC}_i\}+F_t\mathrm{ICR} \tag{7.8}$$

$$S_{pj}+\sum_{k=1}^{K_p}x_{pk}D_{pjk}\le S_{ij},\quad p\in P_i,\ i=1,\ldots,I,\ j=1,\ldots,J \tag{7.9}$$

$$\sum_{j=1}^{J}\left\{y_{ijl}\left(S_{ij}+\sum_{k=1}^{K_i}x_{ik}D_{ijk}+\mathrm{INT}_{ij}\right)-y_{i,j,l+1}S_{ij}\right\}=0, \quad i=1,\ldots,I,\ l=1,\ldots,J-1 \tag{7.10}$$

$$\sum_{k=1}^{K_i}x_{ik}L_{ik}\left(\sum_{j=1}^{J}\mathrm{INT}_{ij}\right)\le\mathrm{IRC}_i,\quad i=1,\ldots,I \tag{7.11}$$

$$S_{Ij} + \sum_{k=1}^{K_I} x_{Ik} D_{Ijk} \le F_t, \quad j = 1, \ldots, J \tag{7.12}$$

$$F_t \le T_{\max} \tag{7.13}$$

$$\sum_{k=1}^{K_i} x_{ik} = 1, \quad i = 1, \ldots, I \tag{7.14}$$

$$\sum_{j=1}^{J} y_{ijl} = 1, \quad i = 1, \ldots, I, \quad l = 1, \ldots, J \tag{7.15}$$

$$\sum_{l=1}^{J} y_{ijl} = 1, \quad i = 1, \ldots, I, \quad j = 1, \ldots, J \tag{7.16}$$

The objective function (7.8) minimizes the project cost TC. Constraints (7.9) guarantee that the precedence relations between activities in each unit are preserved. Constraints (7.10) ensure that every activity is performed according to the given logical sequence. In other words, the sub-activity scheduled in sequence $l+1$ can begin only after the completion of the sub-activity of the same activity scheduled in sequence l. Constraints (7.11) limit the range of idle resource costs for all activities. Constraints (7.12) ensure that the project duration is greater than the finish times of all units in the finish activity. Constraint (7.13) forces the project duration to not exceed the given deadline. Constraints (7.14) require every activity to be performed in only one mode. Constraints (7.15) guarantee that each activity is scheduled in only one sequence. Constraints (7.16) ensure that any two sub-activities of the same activity cannot be performed simultaneously.

7.3.3 Proposed Genetic Algorithm

The above model takes the form of mixed integer nonlinear programming. Exact methods such as nonlinear programming cannot provide an optimal solution for this problem given their NP-hard complexity and the introduction of soft logic. Therefore, the GA technique is applied to solve this problem.

7.3.3.1 Encoding of Chromosomes

The chromosome is designed to represent the two types of decision variables in the DTCTP-*sl*, including the modes and the logical sequences to be selected for all activities. Figure 7.4 shows an example

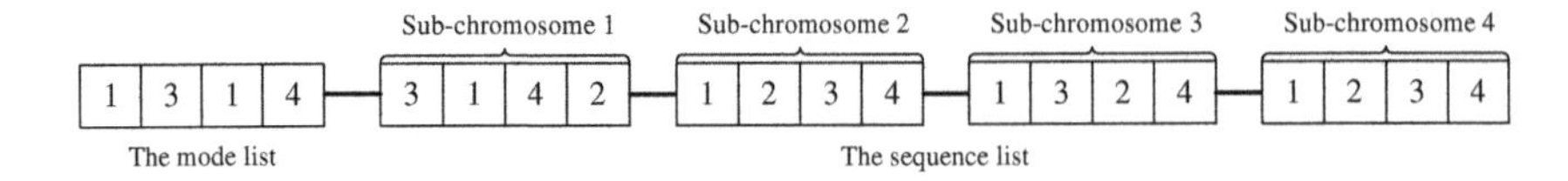

Figure 7.4 The structure of a chromosome.

of the structure of a chromosome. The first list, called the mode list, indicates the modes of all activities. The jth ($j = 1, \ldots, M$) gene in this list is valued by one of the modes of activity i. The second list, called the sequence list, is used to encode the logical sequence of all activities; here, each sub-chromosome is a permutation of all the integers from 1 to J. The value of the jth ($j = 1, \ldots, M$) gene in sub-chromosome i represents the work sequence of a_{ij}.

7.3.3.2 Decoding of Chromosomes

In this section, the proposed GA attempts to construct a project schedule from each chromosome I. First, we calculate the durations of all sub-activities by decoding the model list. Next, a decision set is determined. It includes all the unscheduled activities whose predecessors have been scheduled. An activity is chosen randomly, and according to the logical sequence determined by the sequence list, each sub-activity in this activity is scheduled at its earliest feasible start time. Once all the activities are scheduled, the project duration of chromosome I, denoted by $F_t(I)$, is derived and the corresponding project cost $F_t(I)$ is calculated using Eq. (7.8). Third, if $F_t(I) \leq T_{\max}$, we adopt the model of DTCTP-*sl* to recalculate S_{ij} and INT_{ij} to obtain the minimum project cost of chromosome I. A chromosome implies a mode assignment and logical sequence decision; that is, x_{ik} and y_{ijl} have been determined. Then this model becomes a linear programming model because constraints (7.14)–(7.16) can be removed and the objective function, constraint (7.10), and constraint (7.11) can be linearized.

7.3.3.3 Fitness Computation

In the DTCTP-*sl*, not all the project schedules generated from the chromosomes meet the deadline constraint. The proposed GA employs a punishment mechanism developed by Peng and Wang (2009) for punishing these infeasible chromosomes in order to ensure that they have smaller fitness than the feasible ones. According to the punishment mechanism, the revised objective function of chromosome I, denoted by $F'_c(I)$, is defined as

$$F'_c(I) = (1 + \beta_t)F_c(I) \tag{7.17}$$

where β_t is a positive penalty factor calculated by

$$\beta_t = \begin{cases} \dfrac{F_t(I) - T_{\max}}{\overline{T} - T_{\max}}, & F_t(I) > T_{\max}, \\ 0, & F_t(I) \leq T_{\max}. \end{cases} \tag{7.18}$$

where $\overline{T}$ is the maximal project duration of all the project schedules. Since this is a minimization problem, the fitness function defined in (7.17) is written as

$$f_c(I) = \max_{H=1,\ldots N_P} \{F'_c(H)\} - F'_c(I) \tag{7.19}$$

where N_P denotes the size of the population. After the fitness computation is completed, N_P pairs of chromosomes are randomly selected through roulette selection to undergo the following evolutionary operations.

7.3.3.4 Crossover Operator

The one-point crossover is used for the mode list. For the sequence list, two types of crossover operators are employed. First, we draw a random integer r with $0 \leq r \leq I$ as the crossover point, and the activities $i = 1, 2, \ldots, r$ in the offspring take genes from one parent; that is, $l^i_j(d) = l^i_j(f),\ j = 1, 2, \ldots, J$, where $l^i_j(d)$ and $l^i_j(f)$ are the work sequences of sub-activity a_{ij} in the offspring and father, respectively. Meanwhile, the activities $i = r + 1, \ldots, I$ in the offspring are filled with genes from another parent; that is, $l^i_j(d) = l^i_j(m),\ j = 1, 2, \ldots, J$, where $l^i_j(m)$ is the work sequence of sub-activity a_{ij} in the mother. Next, the position-based crossover operator proposed by Syswerda (1991) is applied in a randomly selected sub-chromosome of the offspring. Specifically, this sub-chromosome takes some genes from one parent at random and fills the vacuum position with genes from another parent through a left-to-right scan.

7.3.3.5 Mutation Operator

The mutation operator is applied on the chromosome generated from the crossover operation with a probability of mutation p_{mut}. For the mode list, the one-point mutation is used in a randomly selected activity i. Next, the value of the gene corresponding to this activity is transformed to another random integer within the range $[1, K_i]$. For the sequence list, we take the following steps to complete the mutation operation: (1) randomly select a sub-chromosome; (2) randomly select two genes in this sub-chromosome; and (3) swap the positions of these two selected genes.

7.3.4 Case Study

Again we take the concrete bridge construction project in Section 7.2.3 for our example. Under hard logic, the minimum project total cost and project direct cost calculated by Hyari et al. (2009) are $1,668,021 and $1,317,642, respectively, and their corresponding project durations are 124 days and 143 days, respectively. In addition, the minimum project duration is 107 days, and the corresponding total cost is $1,736,861.

When soft logic is considered, the minimum project total cost is reduced to $1,618,868, a savings of $117,993, and the corresponding project duration is reduced to 117.34 days. The minimum project direct cost is equal to that where hard logic is assumed, but the corresponding project duration is shortened to 130 days, a saving of 13 days. On the other hand, the shortest project duration where soft logic is considered (94 days) is much less than that where the fixed logical sequence is adopted (107 days), a saving of 12%. Table 7.3 presents all the non-dominated solutions between project duration and direct cost by applying the proposed GA.

Figure 7.5 shows a comparison of complete time–cost profiles based on hard logic and soft logic. This comparison supports the thesis that applying soft logic to scheduling provides more flexibility in reducing project cost and project duration.

7.4 CONCLUSION AND PROSPECTS

This chapter extends the classical DTCTP in repetitive construction projects for two new problems, namely MVDTCTP and DTCTP-*sl*. The MVDTCTP permits all activities to change their execution modes in different units, and in DTCTP-*sl* each activity is allowed to be scheduled in various logical sequences. New nonlinear programming

Table 7.3 Non-Dominated Solutions Calculated by the Proposed GA

Deadline in Days	Project Duration in Days	Project Cost in $	Activity Modes/Logical Sequences
93	—	—	—
94	94.00	1,758,235	[1 1 1 1][1 1 1 1][3 3 3 3][1 1 1 1][1 1 1 1]/[4 2 1 3][4 2 1 3][4 2 1 3][4 2 1 3][4 2 1 3]
95	95.00	1,711,476	[1 1 1 1][1 1 1 1][3 3 3 3][1 1 1 1][1 1 1 1]/[4 3 2 1][4 3 2 1][4 3 2 1][4 3 2 1][4 3 2 1]
96	96.00	1,691,233	[1 1 1 1][1 1 1 1][3 3 3 3][2 2 2 2][1 1 1 1]/[4 3 2 1][4 3 2 1][4 3 2 1][4 3 2 1][4 3 2 1]
97	97.00	1,683,690	[1 1 1 1][1 1 1 1][3 3 3 3][2 2 2 2][1 1 1 1]/[4 3 2 1][4 3 2 1][4 3 2 1][4 3 2 1][4 3 2 1]
98	98.00	1,681,276	[1 1 1 1][1 1 1 1][3 3 3 3][2 2 2 2][1 1 1 1]/[4 3 2 1][4 3 2 1][4 3 2 1][4 3 2 1][4 3 2 1]
99	99.00	1,670,098	[1 1 1 1][1 1 1 1][3 3 3 3][3 3 3 3][1 1 1 1]/[4 3 2 1][4 3 2 1][4 3 2 1][4 3 2 1][4 3 2 1]
100	100.00	1,664,177	[1 1 1 1][1 1 1 1][3 3 3 3][3 3 3 3][1 1 1 1]/[4 2 3 1][4 2 3 1][4 2 3 1][4 2 3 1][4 2 3 1]
101	101.00	1,657,330	[1 1 1 1][1 1 1 1][3 3 3 3][3 3 3 3][1 1 1 1]/[4 2 3 1][4 2 3 1][4 2 3 1][4 2 3 1][4 2 3 1]
102	102.00	1,650,767	[1 1 1 1][2 2 2 2][3 3 3 3][3 3 3 3][1 1 1 1]/[4 3 2 1][4 3 2 1][4 3 2 1][4 3 2 1][4 3 2 1]
103	103.00	1,647,682	[1 1 1 1][2 2 2 2][2 2 2 2][3 3 3 3][1 1 1 1]/[4 2 1 3][4 2 1 3][4 2 1 3][4 2 1 3][4 2 1 3]
104	104.00	1,643,361	[1 1 1 1][2 2 2 2][3 3 3 3][3 3 3 3][1 1 1 1]/[4 1 2 3][4 1 2 3][4 1 2 3][4 1 2 3][4 1 2 3]
105	105.00	1,643,317	[1 1 1 1][2 2 2 2][3 3 3 3][3 3 3 3][1 1 1 1]/[4 1 2 3][4 1 2 3][4 1 2 3][4 1 2 3][4 1 2 3]
106–107	105.30	1,643,304	[1 1 1 1][2 2 2 2][3 3 3 3][3 3 3 3][1 1 1 1]/[4 1 2 3][4 1 2 3][4 1 2 3][4 1 2 3][4 1 2 3]
108	108.00	1,639,468	[1 1 1 1][2 2 2 2][3 3 3 3][4 4 4 4][1 1 1 1]/[4 1 3 2][4 1 3 2][4 1 3 2][4 1 3 2][4 1 3 2]
109–110	108.60	1,638,786	[1 1 1 1][2 2 2 2][3 3 3 3][4 4 4 4][1 1 1 1]/[4 1 2 3][4 1 2 3][4 1 2 3][4 1 2 3][4 1 2 3]
111–117	110.95	1,635,388	[1 1 1 1][2 2 2 2][2 2 2 2][4 4 4 4][1 1 1 1]/[4 1 3 2][4 1 3 2][4 1 3 2][4 1 3 2][4 1 3 2]
118 –	117.34	1,618,868	[1 1 1 1][3 3 3 3][1 1 1 1][4 4 4 4][1 1 1 1]/[4 3 1 2][4 3 1 2][4 3 1 2][4 3 1 2][4 3 1 2]

model formulations and GA-based optimization procedures are presented for both problems. In comparison with existing studies, the proposed models can increase the number of feasible solutions and may expand the range of values of optimization criteria. This is important for both owners and contractors.

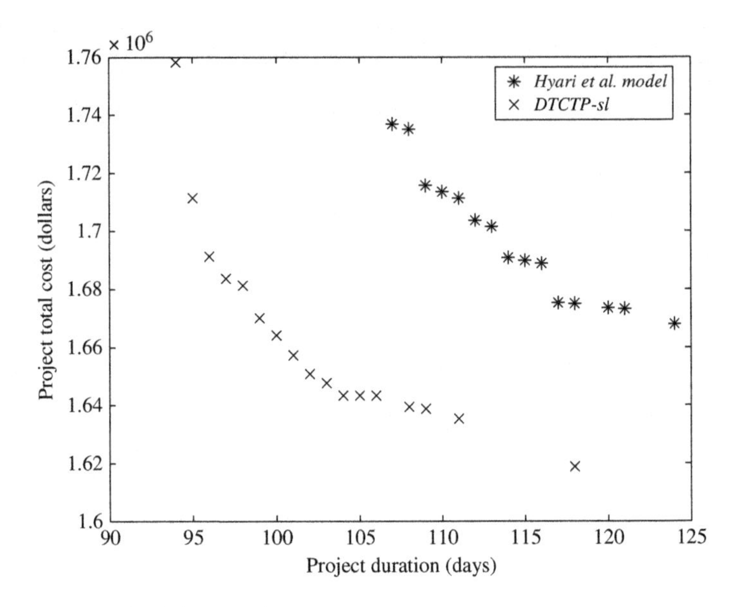

Figure 7.5 Comparison of calculation results.

An important assumption adopted in the DTCTP-*sl* is that all units of each activity can only be scheduled in sequence. In other words, for those projects that employ more than one crew to perform the same activity, the proposed model has only a reference meaning. This is the main limitation of the model, and we will attempt to improve it in our future studies. On the other hand, we will continue our studies to analyze the applications of soft logic to other problems such as resource leveling and resource-constrained project scheduling.

REFERENCES

Ahuja, R.K., Magnanti, T.L., Orlin, J.B., 1989. Network flows. In: Nelhauser, G.L., Rinnooy Kan, A.H.G., Todd, M.J. (Eds.), Handbooks in Operation Research and Management Science. Elsevier, Amsterdam, pp. 258–263.

Al Sarraj, Z.M., 1990. Formal development of line-of-balance technique. J. Constr. Eng. Manage. 116 (4), 689–704.

Ammar, M.A., 2013. LOB and CPM integrated method for scheduling repetitive projects. J. Constr. Eng. Manage. 139 (1), 44–50.

Ammar, M.A., Elbeltagi, E., 2001. Algorithm for determining controlling path considering resource continuity. J. Comput. Civ. Eng. 15 (4), 292–298.

Ammar, M.A., Mohieldin, Y., 2002. CPM/RSM: CPM-based repetitive scheduling method. III Middle East Regional Civil Engineering Conference, Egypt Section. ASCE, Reston, VA.

Arditi, D., Albulak, M.Z., 1986. Line-of-balance scheduling in pavement construction. J. Constr. Eng. Manage. 112 (3), 411–424.

Arditi, D., Tokdemir, O.B., Suh, K., 2002. Challenges in line-of-balance scheduling. J. Constr. Eng. Manage. 128 (6), 545–556.

Badiru, A.B., 1992. Computational survey of univariate and multivariate learning curve models. IEEE Trans. Eng. Manage. 39 (2), 176–188.

Blazewicz, J., Lenstra, J., Rinnooy Kan, A., 1983. Scheduling subject to resource constraints: classification and complexity. Discrete Appl. Math. 5, 11–24.

Brucker, P., Drexl, A., Möring, R., Neumann, K., Pesch, E., 1999. Resource constrained project scheduling: notation, classification, models, and methods. Eur. J. Oper. Res. 112, 3–41.

Carr, R.I., Meyer, W.L., 1974. Planning construction of repetitive building units. J. Constr. Div. 100 (3), 403–412.

Chrzanowski Jr., E.N., Johnston, D.W., 1986. Application of linear scheduling. J. Constr. Eng. Manage. 112 (4), 476–491.

Demeulemeester, E.L., Herroelen, W., 2002. Project scheduling-A research handbook. Volume 49 of International Series in Operations Research & Management Science. Kluwer Academic Publishers, Boston, MA.

Elloumi, S., Fortemps, P., 2010. A hybird rank-based evolutionary algorithm applied to multi-mode resource-constrained project scheduling problem. Eur. J. Oper. Res. 205, 31–41.

Elmaghraby, S.E., Kamburowski, J., 1992. The analysis of activity networks under generalized precedence relations (GPRs). Manage. Sci. 38 (9), 1245–1263.

El-Rayes, K., Moselhi, O., 2001. Optimizing resource utilization for repetitive construction projects. J. Constr. Eng. Manage. 127 (1), 18–27.

El-Sersy, A.H., 1992. An Intelligent Data Model for Schedule Updating (Ph.D. dissertation). University of California, Berkeley, CA.

Ezeldin, A.S., Soliman, A., 2009. Hybird time-cost optimization of nonserial repetitive construction projects. J. Constr. Eng. Manage. 135 (1), 42–55.

Fan, S.L., Tserng, H.P., 2006. Object-oriented scheduling for repetitive projects with soft logics. J. Constr. Eng. Manage. 132, 35–48.

Fan, S.L., Sun, K.S., Wang, Y.R., 2012. GA optimization model for repetitive projects with soft logic. Autom. Constr. 21 (1), 253–261.

Gen, M., Cheng, R., 2000. Genetic Algorithm and Engineering Optimization. John Wiley and Sons, New York, NY.

Gransberg, D.D., 2007. Converting linear schedules to critical path method precedence. AACE Int. Trans. 5, 1–4.

Harmelink, D.J., Rowings, J.E., 1998. Linear scheduling model: development of controlling activity path. J. Constr. Eng. Manage. 124 (4), 263–268.

Harris, R.B., Ioannou, P.G., 1998. Scheduling projects with repeating activities. J. Constr. Eng. Manage. 124 (4), 269–278.

Hartmann, S., Kolisch, R., 2000. Experimental evaluation of state-of-the-art heuristics for the resource-constrained project scheduling problem. Eur. J. Oper. Res. 127, 307–394.

Hegazy, T., Wassef, N., 2001. Cost optimization in projects with repetitive non-serial activities. J. Constr. Eng. Manage. 127 (3), 183–191.

Hegazy, T., 2002. Critical path method – line of balance model for efficient scheduling of repetitive construction projects. Transp. Res. Rec. 1761, 124–129.

Herroelen, W., De Reyck, B., Demeulemeester, E., 1998. Resource-constrained project scheduling: a survey of recent development. Comput. Oper. Res. 25, 279–302.

Hsie, M., Chang, C.J., Yang, I.T., Huang, C.Y., 2009. Resource-constrained scheduling for continuous repetitive projects with time-based production units. Autom. Constr. 18, 942–949.

Hyari, K., El-Rayes, K., 2006. Optimal planning and scheduling for repetitive construction projects. J. Manage. Eng. 22 (1), 11–19.

Hyari, K.H., El-Rayes, K., El-Mashaleh, M., 2009. Automated trade-off between time and cost in planning repetitive construction projects. Constr. Manage. Econ. 27 (8), 749–761.

Icmeli, O., Erenguc, S., Zappe, C., 1993. Project scheduling problems: A survey. Int. J. Oper. Prod. Manage. 13, 80–91.

Ipsilandis, P.G., 2007. Multiobjective linear programming model for scheduling linear repetitive projects. J. Constr. Eng. Manage. 133 (6), 417–424.

Jarkas, A.M., 2010. Critical investigation into the applicability of the learning curve theory to rebar fixing labor productivity. J. Constr. Eng. Manage. 136 (12), 1279–1288.

Johnston, D.W., 1981. Linear scheduling method for highway construction. J. Constr. Eng. Manage. 107 (CO_2), 247–260.

Kallantzis, A., Lambropoulos, S., 2004. Critical path determination by incorporating minimum and maximum time and distance constraints into linear scheduling. Eng. Constr. Archit. Manage. 11 (3), 211–222.

Kallantzis, A., Soldatos, J., Lambropoulos, S., 2007. Linear versus network scheduling: a critical path comparison. J. Constr. Eng. Manage. 133 (7), 483–491.

Kang, L.S., Park, I.C., Lee, B.H., 2001. Optimal schedule planning for multiple, repetitive construction process. J. Constr. Eng. Manage. 127 (5), 382–390.

Kolisch, R., Hartmann, S., 2006. Experimental investigation of heuristics for resource-constrained project scheduling: an update. Eur. J. Oper. Res. 174, 23–37.

Lam, K.C., Lee, D., Hu, T., 2001. Understanding the effect of the learning-forgetting phenomenon to duration of projects construction. Int. J. Project Manage. 19 (7), 411–420.

Leu, S.S., Hwang, S.T., 2001. Optimal repetitive scheduling model with shareable resource constraint. J. Constr. Eng. Manage. 127 (4), 270–280.

Liu, S.S., Wang, C.J., 2007. Optimization model for resource assignment problems of linear construction projects. Autom. Constr. 16, 460–473.

Long, L.D., Ohsato, A., 2009. A genetic algorithm-based method for scheduling repetitive construction projects. Autom. Constr. 18 (4), 499–511.

Lucko, G., 2009. Productivity scheduling method: linear schedule analysis with singularity functions. J. Constr. Eng. Manage. 135 (4), 246–253.

Lumsden, P., 1968. The Line-of-Balance Method. Pergamon, Tarrytown, NY.

O'Brien, J.J., 1975. VPM scheduling for high rise buildings. J. Constr. Div. 101 (4), 895–905.

Pellegrino, R., Costantino, N., Pietroforte, R., Sancilio, S., 2012. Construction of multi-storey concrete structures in Italy: patterns of productivity and learning curves. Constr. Manage. Econ. 30 (2), 103–115.

Peng, W.L., Wang, C.G., 2009. A multi-mode resource-constrained discrete time-cost trade/off problem and its genetic algorithm based solution. Int. J. Project Manage. 27, 600–609.

Reda, R.M., 1990. RPM: repetitive project modeling. J. Constr. Eng. Manage. 116 (2), 316–330.

Russell, A., Caselton, W., 1988. Extensions to linear scheduling optimization. J. Constr. Eng. Manage. 114 (1), 36–52.

Selinger, S., 1980. Construction planning for linear projects. J. Constr. Div. 106 (2), 195–205.

Senior, B.A., 1993. A Study of the Planning and Integrated Cyclic Analysis of Serial System Operations (Ph.D. Thesis). Purdue University, West Lafayette, IN.

Senouci, A.B., Eldin, N.N., 1996. Dynamic programming approach to scheduling of nonserial linear project. J. Comput. Civ. Eng. 10 (2), 106–114.

Slowinski, R., 1980. Two approaches to problems of resource allocation among project activities – a comparative study. J. Oper. Res. Soc. 31 (8), 711–723.

Suhail, S.A., Neale, R.H., 1994. CPM/LOB: new methodology to integrate CPM and line of balance. J. Constr. Eng. Manage. 120 (3), 667–684.

Syswerda, G., 2001. Schedule optimization using genetic algorithms. In: Davis, L. (Ed.), Handbook of Genetic Algorithms. Van Nostrand Reinhold, New York, NY.

Talbot, F., 1982. Resource-constrained project scheduling problem with time-resource trade-offs: the nonpreemptive case. Manage. Sci. 28, 1197–1210.

Tamimi, S., Diekmann, J., 1988. Soft logic in network analysis. J. Comput. Civ. Eng. 2 (3), 289–300.

Terry, S.B., Lucko, G., 2012. Algorithm for time-cost trade/off analysis in construction projects by aggregating activity-level singularity functions. Proceedings of the 2012 Construction Research Congress.

Thabet, W.Y., Beliveau, Y.J., 1994. HVLS: horizontal and vertical logic scheduling for multistory projects. J. Constr. Eng. Manage. 120 (4), 875–892.

Vanhoucke, M., 2006. Work continuity constraints in project scheduling. J. Constr. Eng. Manage. 132 (1), 14–25.

Vorester, M.C., Beliveau, Y.J., Bafna, T., 1992. Linear scheduling and visualization. Transp. Res. Rec. 1351, 32–39.

Wang, C.H., Huang, Y.C., 1998. Controlling activity interval times in LOB scheduling. Constr. Manage. Econ. 16 (1), 5–16.

Wang, W.C., 2005. Impact of soft logic on the probabilistic duration of construction projects. Autom. Constr. 23, 600–610.

Wittrick, W.H., 1965. A generalization of Macaulay's method with applications in structural mechanics. AIAA J. 3 (2), 326–330.

Yamin, R.A., Harmelink, D.J., 2001. Comparison of linear scheduling model (LSM) and critical path method (CPM). J. Constr. Eng. Manage. 127 (5), 374–381.

Yang, I., 2002. Repetitive Project Planner Resource-Driven Scheduling for Repetitive Construction Projects (Ph.D. Dissertation). University of Michigan, Ann Arbor, MI.

Yang, T., Ioannou, P., 2004. Scheduling system with focus on practical concerns in repetitive projects. J. Constr. Eng. Manage. 22 (6), 619–630.

Zhang, L.H., Qi, J.X., 2012. Controlling path and controlling segment analysis in repetitive scheduling method. J. Constr. Eng. Manage. 138 (11), 1341–1345.

Printed in the United States
By Bookmasters